Grasslands around the world provide most of the world's milk and wool
and much of the meat. The growing world population and the threat of
global warming have put increased pressure on the world's grasslands.

Ross Humphreys has played a leading role in the development of
grassland science and in this authoritative review discusses how it has
evolved over the past 60 years.

Using the proceedings of the International Grassland Congress since
1937, Professor Humphreys identifies the shifting emphasis of the science,
assesses the current state of play and looks at perspectives for the future.

This unique review will be of great value to all grassland scientists and
ecologists.

THE EVOLVING SCIENCE OF GRASSLAND IMPROVEMENT

THE EVOLVING SCIENCE OF GRASSLAND IMPROVEMENT

L.R. HUMPHREYS

Emeritus Professor of Agriculture,
University of Queensland,
St Lucia, Australia

CAMBRIDGE
UNIVERSITY PRESS

CAMBRIDGE UNIVERSITY PRESS
Cambridge, New York, Melbourne, Madrid, Cape Town, Singapore, São Paulo

Cambridge University Press
The Edinburgh Building, Cambridge CB2 8RU, UK

Published in the United States of America by Cambridge University Press, New York

www.cambridge.org
Information on this title: www.cambridge.org/9780521495677

© Cambridge University Press 1997

This publication is in copyright. Subject to statutory exception
and to the provisions of relevant collective licensing agreements,
no reproduction of any part may take place without the written
permission of Cambridge University Press.

First published 1997
This digitally printed version 2007

A catalogue record for this publication is available from the British Library

Library of Congress Cataloguing in Publication data

Humphreys, L. R.
    The evolving science of grassland improvement / L. R. Humphreys.
        p.   cm.
    Includes bibliographical references (p.   ) and index.
    ISBN 0 521 49567 9
    1. Pastures.   2. Forage plants.   3. Grasslands.   4. Pasture
ecology.   5. Grassland ecology.   I. Title.
    SB199.H84   1997
    633.2′02–dc20   96-9419 CIP

ISBN 978-0-521-49567-7 hardback
ISBN 978-0-521-03873-7 paperback

To I.L.H.

Contents

Preface

Grasslands provide most of the world's milk and wool and much of the meat; they are also integrated in many regions with the production of annual and perennial crops. The threat of global warming accords a special premium to the carbon sequestered by grasslands. The rising expectations of the increasing global population place pressure on grasslands and on marginal lands developed for cropping. The greater demand for grassland products has been met in part by farmers applying the science of grassland improvement.

Grassland science has made notable advances in recent decades and as a holistic science its progress has been marked by changes in the emphasis given to its component disciplines, which have each made differing gains. The improvement of grassland rests upon plant and soil sciences, animal nutrition, systems theory, which generates decision support systems for management, and socio-economic perspectives of the role of the farmer and of institutional policies.

The International Grassland Congress, which has met on 17 occasions since its first meeting in 1927, has contributed to this evolution. The proceedings of its recent meetings provide markers for understanding the contemporary condition of grassland science and its probable future directions, whilst its earlier meetings indicate the origins and history of the major changes of emphasis and of the controversies in which scientists have been embroiled regarding grassland improvement and management.

These differences of viewpoint, often allied to regional needs or national imperatives, include the emerging concerns for environmental protection which challenge the drive for increased production and the attitudes to the balance of grassland, crop land and woodland. The opening chapter discusses different styles of grassland improvement and the impact of grassland on greenhouse gases.

There follow three major themes basic to grassland improvement. The recognition of elite plant germplasm and the constitution of new gene combinations help to optimize the supply of digestible herbage acceptable to livestock and to protect environmental resources from degradation. The debate about suitable criteria of plant merit, the conservation and global sharing of germplasm, the reliance on plant introduction or plant breeding for plant improvement and the emerging role of molecular biology is presented. Plant introduction may result in the spread of pests and conflicts with the preservation of 'natural' grasslands. The nitrogen economy of grasslands is then chosen as the major edaphic theme, since nitrogen availability is the primary mineral constraint to plant and animal production from grassland; the reliance on the legume for biological nitrogen fixation or on nitrogen fertilizer produced from fossil fuel, and the atmospheric and stream pollution arising from nitrogen in agricultural systems, are controversial. The third strand is the manipulation of the leaf surface by defoliation, which optimizes the sustained harvesting of forage nutrients and maintains protective cover of the soil. The early controversies about the role of carbohydrate 'reserves' and the leaf area index are reworked in terms of carbon and nitrogen allocation to shoot meristems.

The book then treats the interplay of biotic, climatic and edaphic components of the environment as these modify the dynamics of grassland communities. The demise of linear Clementsian succession, the rise of the state-and-transition model, the replacement of the evangelistic piety of range management by more rational and quantitative assessments of land condition, and the advances in describing processes of population change are discussed. This leads on to issues of grazing management, which are directed to the synchrony of the supply of available forage with the demand of animals grazing the pasture and to the maintenance of the vigour of acceptable pasture. Stocking rate, forage allowance and sward surface height are the dominant considerations, and the disadvantages of rotational and 'time control' grazing systems are set against alternative approaches of seasonal adjustment in grazing pressure.

The final chapter indicates how the former dependence on reductionist science has been complemented by systems and modelling theory, whilst the concept of the grassland scientist has been enlarged by interaction with sociologists and economists concerned with equity in grassland development. The search for innovation has been balanced by efforts to optimize processes using the approach of hard systems; the inadequacy of these approaches has led to work with soft systems directed to achieving the fulfilment of the goals of farmers and communities.

An appendix sets out the venues and varying regional attendance at the International Grassland Congress, its organizational evolution and the changes in the balance of disciplinary themes discussed.

The above outline indicates considerable selectivity of topics. Originally I set out 130 topics grouped into ten major themes. I then noted that the proceedings of the 1993 XVII International Grassland Congress alone weigh 6 kg, and that it would be best to focus on significant themes which had aroused controversy and in which scientific emphases had changed, and to avoid areas of the worst personal ignorance. I was also influenced by the need not to recapitulate in an egregious way the predominant themes of my writing in recent years about tropical pastures: the integration of forages in cropping systems, seed production and pasture utilization. However, a few illustrations have crept in from Humphreys (1981), which went out of print in 1985. There is more material about ideas and research results than about techniques. The worst gaps in the book's treatment of grassland improvement relate to forage nutritive value, which gets a mention in chapter 2, ruminant nutrition, and approaches to maintaining continuity of forage supply, especially through forage processing. These topics are well covered by Hacker & Ternouth (1987), de Boer & Bickel (1988), Garnsworthy (1988), Minson (1990), Jung *et al.* (1993) and Fahey *et al.* (1994).

The first International Grassland Congress I attended was held at Reading, UK in 1960; it affected profoundly my professional orientation. This book has been written partly to assist young scientists in developing perspectives about the ground from which the component disciplines of grassland science have emerged and interacted, and which now shape its future directions.

L.R. Humphreys
University of Queensland
March, 1996

Acknowledgements

I am indebted to B.W. Norton and H.M. Shelton, Department of Agriculture, University of Queensland, Sir David Smith FRS FRSE, Wolfson College, University of Oxford and C.J. Leaver FRS FRSE, Department of Plant Sciences, University of Oxford for support and to Carolyn Smith for typing the manuscript. Special assistance in the preparation of this book was given by D.F. Cameron, R.A. Date, R.M. Jones, J.G. McIvor and J.R. Wilson of the CSIRO Division of Tropical Crops and Pastures, G.M. McKeon of the Queensland Department of Primary Industries, and R.L. Ison of the Open University, Milton Keynes, but I am responsible for the views expressed.

I am also grateful for advice or assistance from W.H. Burrows, S. Chakraborty, R.J. Clements, P.G. Cox, M.J. Fisher, A.S. Foot, J.Z. Foot, J.B. Hacker, R.L. Haggar, R.L. Hall, A. Hentgen, T.J.V. Higgins, S.M. Howden, S.C. Jarvis, O.R. Jewiss, Q.R. Jones, R.M. Jones, J.K. Leslie, A. Litherland, L. t'Mannetje, M. Murphy, C.J. Pearson, S. Petty, A.B. Pittock, D.P. Poppi, M. Stafford Smith, B. Walker, A. Weir, R.J. Wilkins, B. Willitts and W.H. Winter.

Abbreviations

AU	Animal unit
b	beasts
CIAT	Centro Internacional Agricultura Tropical
CP	Crude protein
CSIRO	Commonwealth Scientific and Industrial Research Organization
D	Digestibility
DM	Dry matter
DSS	Decision support system
FSR	Farming systems research
hd	head
IGC	International Grassland Congress
IVD DM	*In vitro* digestible dry matter
IVDOM	*In vitro* digestible organic matter
IPCC	Intergovernmental Panel on Climate Change
LAI	Leaf area index
LW	Live weight
LWG	Liveweight gain
ME	Metabolizable energy
NPP	Net primary productivity
OM	Organic matter
ppm	parts per million
RGR	Relative growth rate
SR	Stocking rate
SSH	Sward surface height
TDN	Total digestible nutrients
TNC	Total non-structural carbohydrate
UME	Utilized metabolizable energy
VFA	Volatile fatty acids

1

Grassland improvement and environmental protection

1.1 The science of grassland improvement

Grassland occupies $c.$ 25% of the global land surface (Mooney, 1993) and is central to human welfare. Our understanding of the functioning of grassland and the ways in which it may be modified to meet particular objectives has evolved discontinuously during the last 60 years as scientists made imaginative leaps in grasping the processes at work, validated their insights (or not), and altered the directions of their research in response to inputs from farmers, societal pressures and changing fashion. A good benchmark is Stapledon (1933). This book selectively reviews aspects of the subsequent evolution and forecasts future directions.

Grassland 'improvement' denotes an advance upon a previous condition. An advance may refer to increased efficiency of production, which has been a dominant theme at all of the International Grassland Congresses since the first was held in 1927; it has been balanced by other emerging emphases in recent decades, such as the conservation of resources and the reduction of off-farm effects of agricultural development. Grassland improvement is a necessary objective for native rangelands, of which $c.$ 75% in Africa, South America and Asia and $c.$ 55% in Australia are regarded as 'degraded' (UNEP, 1991). The concept of grassland improvement has come to reflect wider metaphors of agriculture (McClintock & Ison, 1994) which include countryside design, and 'Every landscape has an equally appropriate use, given a different set of values' (Foran, 1993).

Advances in grassland science depend upon a wide base of learning in the plant, soil and animal sciences (which are underpinned by the physical sciences), and in the social sciences. Some of the central ideas revolve about:

(i) The recognition of elite plant germplasm and the constitution of new gene combinations. These are plants which optimize the

conversion of carbon dioxide, water and minerals to supply
digestible herbage acceptable to livestock and which protect
environmental resources from degradation by maintaining soil cover
and accumulating organic matter.

(ii) The enhancement of soil quality (Lal & Miller, 1993). Nitrogen
availability is the primary mineral constraint to plant and animal
production from grassland. Much grassland science deals with
biological N fixation, appropriate use of N fertilizer, the supply of
other nutrients which limit N response, and the efficient cycling of
nutrients which minimizes losses to groundwater, streams, and the
atmosphere. Salinity, mineral toxicities, the stability and structure of
the soil and its biological activity are other primary edaphic themes.

(iii) The manipulation of the leaf surface by management of defoliation
which optimizes the sustained harvesting of herbage nutrients
consistent with continuing soil cover. Pathways of photosynthesis,
resistance to environmental stresses, the control of flowering and
seed production, tillering and plant regeneration are also key areas
of plant physiology.

(iv) Grassland ecology as the interplay of the biotic, climatic and
edaphic components of the environment which modify the dynamics
of grassland communities. This provides conceptual bases for the
management of natural plant communities and for manipulating the
botanical composition of planted pastures, especially as directed to
the maintenance of legume components or to the control of weeds.

(v) Grazing management directed to synchronizing the supply of available
forage with the demand of the animals grazing the pasture and to
maintaining the vigour of acceptable pasture. This overlaps the
physiological theme of sward response to defoliation mentioned above,
but has a stronger focus on the animal response to forage availability.

(vi) Continuity of forage supply. This is attained by the provision of a
sequence of different forages, the modification of the environment in
which the pasture grows, the adjustment of animal requirement to
the forage available and the processing and conservation of feeds.
Feeding systems are developed in which product output reflects
market demand, the genetic capacity of grazing animals and the
nutritive value of available forages as supplemented.

(vii) The realization of change in grasslands which arises from the
participatory rural action of scientists, farmers and resource
managers in formulating problems and accommodating to pathways
of amelioration.

1.2 Functions and properties of pasture systems

1.2.1 Production from grasslands

The key function of product output from grasslands arises from the animals which graze it. The conventional products which are traded, and which also contribute to the direct food security of farm families, are meat, milk, wool, skins and hair. Grassland improvement which results in enhanced quality of diet and less discontinuity of forage supply provides the opportunity to develop higher value products: quality meat from younger carcasses, and milk for sale in the dry or cold season.

Additional products in the less developed economies include animal manure, which may be dried for fuel or conserved for use on crop land (Probert, Okalebo & Jones, 1995), and animal blood. The success of farming systems using animals for draught and carriage may depend on the health and condition of these animals, especially when this determines the rapidity with which an area may be prepared for crop planting at the beginning of a wet season.

Animals raised on grassland and crop products often occupy a key role in the economy of smallholder agriculture. Diversification of income contributes to its stability, and animals constitute an asset readily realizable in times of family crisis or opportunity. They are viewed in some communities as an alternative banking system which hedges savings against inflation and may also be exchanged for grain. Animals meet cultural needs and contribute to farmer status, but these factors may have been overvalued by some sociologists; the situation in subsistence agriculture is rather that animals provide food security for families.

1.2.2 Production from forages in integrated cropping systems

The same products as mentioned above are available from forages in integrated cropping systems, but there are additional benefits which may sustain the cropping system. These are related primarily to the maintenance of soil quality.

The benefits from growing forage legumes either as companions to annual crops (Rerkasem & Rerkasem, 1988) or in rotation with annual crops as a pasture ley arise mainly from biological N fixation (Clatworthy, 1986; Gibson, 1987). This is linked to enhanced organic matter accumulation, which provides benefits to crop production as a nutrient source, through its alleviation of mineral toxicity and increased ion exchange capacity, as an energy source for biological activity, and through effects on soil physical attributes and the deactivation of agricultural chemicals (Humphreys, 1994).

The change in soil physical attributes due to pastures results in increased infiltration, often associated with greater porosity and reduced bulk density (Bridge *et al.*, 1983), and the creation of water stable aggregates. Pastures which are managed to preserve soil cover protect the soil against water and wind erosion (Lal, 1990). This is not a new emphasis; soil conservation was a major theme at the VI International Grassland Congress at State College, Pennsylvania. P.V. Cardon (1952) advocated 'measures aim(ed) at . . . the conquest of factors inimical to soil conservation and sustained productivity of the soil'. On sloping lands a ley system in which the pasture phase predominates may make the long-term rate of soil loss tolerable, or grass strips or contour hedge barriers may provide adequate protection.

Other benefits from the incorporation of forages in cropping systems are due to the more efficient capture of natural resources for plant growth and to enhanced crop protection, as reflected in reduced incidence of weeds (Littler, 1984), pests (Shepherd & Coombs, 1979) and disease (Keswani & Mreta, 1980).

Silvopastoral systems (Copland, Djajanegra & Sabrani, 1994) have gained increasing recognition for their capacity to stabilize landscapes and provide a flexible and diverse income base for farmers. Plantation or timber crops may be combined with herbage production, or shrub legumes (Gutteridge & Shelton, 1994; Shelton, Piggin & Brewbaker, 1995) grown with annual crops are used in varying agroforestry configurations.

1.2.3 Social utility of grasslands

Grassland provides social utility additional to the production systems mentioned above.

Site stabilization

Water is the key product output from many grassland watersheds. This has been a perennial emphasis at many International Grassland Congresses (for instance, Wilm, 1952), and water quality, reliability of flow and yield are modified by the management of watershed cover, whilst the life of water storages is negatively associated with the degree of soil erosion. There are many disturbed situations, especially in relation to mining (McGinnies & Nicholas, 1983), road building, and airfield construction, where special techniques of pasture establishment have been developed.

Turf

The improvement, establishment and management of turf for recreational purposes such as lawns, playing fields and galloping course have become a significant scientific industry in which resistance to trampling and aspects of sward structure are accorded special merit.

Nature conservancy

The increasing urbanization of populations has enhanced the need for rural recreation (Edwards, 1983). This issue may be viewed as distinct from the need for conservation of germplasm and the preservation of relict areas, which is mentioned in chapter 2. R.G. Stapledon (1944) was a strong advocate of the expansion of national parks in the UK, and believed that people's psychological equilibrium could only be maintained through contact with nature. He affirmed that 'land must be considered in relation to the nation as a whole . . . We have to take into account not only the relation of a prosperous countryside . . . to the national well-being, but also the question of providing facilities in the country and amidst truly rural surroundings for the recreation and relaxation of the urban population'.

Stapledon's emphasis on the need to intensify agriculture within the national park was incompatible with the alternative view that 'The conservationists' objective is to make one blade of grass grow where two grew before' (Green, 1990). The design of landscape requires that a consensus be developed about community objectives. In the absence of animals the UK countryside would be impenetrable and unattractive rather than idyllic, and it is questionable whether unforeseen climax vegetation would be aesthetically acceptable and appropriate for leisure activities (Institute of Grassland and Environmental Research, 1992, 1993). Perhaps the community needs are for a diversity of landscapes which embrace both open space grassland and secondary forest.

Management options are illustrated (Table 1.1) for lowlands in the UK (Wilkins & Harvey, 1994). Increased soil fertility is a major factor in increasing farm output and in decreasing the diversity of plant species. Swards are unlikely to have nature conservation value until species density exceeds 15–20 m^{-2}, but even low levels of N input or the build-up of P from previous fertilizer application result in many UK swards having a species density of 1–5 m^{-2}. High levels of stocking rate (SR) reduce the incidence of birds, butterflies and small mammals. The proportion of pastures used for silage rather than hay has increased in the UK (Hopkins *et al.*, 1988);

Table 1.1 *Generalized effects of agricultural management options on utilized output and on wildlife in the UK lowlands*

	Utilized output	No. higher plants	Insects	Birds	Small mammals
Increase fertility	+	−	−	−	0
Increase level of utilization[a]	+	(−)	−	−	−
Increase proportion reseeded	(+)	−	−	−	(−)
Improve drainage	(+)	−	−	−	?
Use herbicides	(+)	−	−	V	?
Use pesticides	(+)	0	−	V	(−)
Increase proportion early-cut	(+)	−	−	−	(−)
Increase proportion late-cut	(−)	+	+	+	+
Graze with cattle rather than sheep	(−)	(+)	+	+	+
Graze intermittently rather continuously	0	V	+	+	V

Note:
The direction of change, either positive or negative is indicated with symbols. Those not in parentheses indicate major effects. V signifies that direction of effect varies in different situations.
[a] Particularly increased severity of grazing.
Source: Wilkins & Harvey (1994).

this, together with more frequent cutting, leads to a loss of ground nesting birds such as the corncrake (*Crex crex*) (Green, 1990).

Wilkins & Harvey (1994) cite studies which indicate that in UK lowland conditions the output from grassland (as Utilized Metabolizable Energy, UME) is not greatly increased by reseeding (7% over 5 years), drainage (7% over 10 years), and herbicides (2%), whilst pesticides are little used. On the other hand, these measures (together with sheep grazing) all diminish nature conservation value. They suggest that the application of contrasting managements to particular fields within a farm is preferable to seeking to marry production and nature conservation objectives across the whole farm.

Control of pollution

This is discussed subsequently, in sections 1.3, 3.43 and 3.44.

1.2.4 *Properties of grassland systems*

Grassland systems are discussed in chapter 7 and incidentally in other chapters. Conway (1994) suggests that agroecosystems be evaluated in

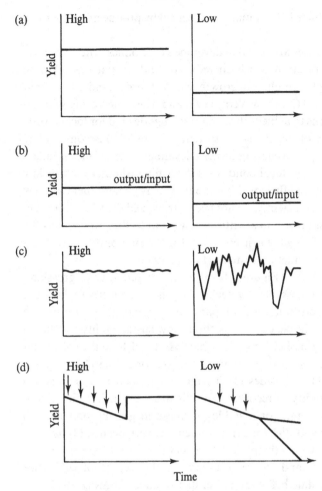

Figure 1.1. The properties of grassland systems of high (left-hand side) and low (right-hand side) (a) productivity, (b) efficiency, (c) stability and (d) sustainability when stress (arrows) is imposed. (From Pearson & Ison, 1987.)

terms of a primary goal of 'social value', measured by the outputs of goods and services, the degree to which these satisfy human needs and their allocation among the human population (equitability). A high level of productivity (Figure 1.1a, Pearson & Ison, 1987) was endorsed as a goal at many International Grassland Congresses, but the emphasis on high rates of fertilizer N application, as discussed in chapter 3, has been moderated by concerns over environmental pollution, the use of fossil fuel, the drain on foreign exchange for countries lacking oil, the subsidization of animal production by other sectors of the economy, and diminished success in placing animal products in freer markers accessed by producers using lower input systems.

Contrasting emphases are found in developed economies, where the level of grassland intensification has been reduced, and in many developing economies with high population growth where the demand for animal products has increased (Nores & Vera, 1993) and where increasing production is seen as a means to achieve the capital savings desired for local investment. The efficiency of the grassland system (Figure 1.1b) becomes a key issue as product competitiveness becomes relevant and inefficiency is linked to nutrient drainage, but local conditions determine whether high or low input systems are more efficient. Low input systems, where there are few nutrients to be cycled efficiently, do not necessarily yield the best return. On the other hand most grassland scientists view grain feeding of cattle as the least energetically efficient, most costly and most environmentally damaging of all animal production systems (Foran, 1993).

The degree of stability (Figure 1.1c) is a further property of grassland systems and the riskiness of new technology is of concern to farmers, according to their attitude, and of paramount concern to subsistence farmers. There is no simple consensus about how sustainability is defined; the time course of grassland output, which is linked to soil quality and sward composition, and its recovery from adverse conditions (Figure 1.1d) is a key concept. This includes (1) inertia, or resistance to change, (2) elasticity, or the rapidity of recovery after disturbance, such as fire, flood, drought, disease epidemic, or a sudden increase in grazing pressure, (3) amplitude, or the size of the zone from which recovery occurs, (4) hysteresis, or the degree to which the path of recovery is the exact reversal of the path of disturbance, and (5) malleability, or the degree of difference between the system state before and after disturbance (Conway, 1994).

1.3 Atmospheric pollution and global warming

The intensification of land use, industrial activity and the concentration of human population in urban areas lead to increased atmospheric pollution. This may be expressed in the production of acid rain, associated especially with increased levels of sulphur (S), enhanced levels of lead, uncongenial particulate pollution related to fly-ash or smoke nuisance and a thinning of the tropospheric ozone layer and reduced entrapment of ultra-violet radiation, although ozone may itself be a plant pollutant (World Commission on Environment and Development, 1987). The suggestion at the IV International Grassland Congress (Vezzani & Carbone, 1937) that the success of mountain grazing is due to 'pure air, rich in ultra-violet rays' is now discounted. The burning of fossil fuels and agricultural production are responsible for raising the atmospheric concentration of the 'greenhouse' gases carbon dioxide (CO_2), methane (CH_4) and nitrous oxide (N_2O) and these changes are expected to increase global temperature and to change regional rainfall and sea levels (Gates, 1993; Pittock, 1993). This segment of the book focuses on the theme of global warming, since the management of grassland can contribute to the amelioration of these effects through the sequestration of carbon (C) or can exacerbate the problem. This topic was merely mentioned at the XVI International Grassland Congress (Humphreys, 1989), but emerged as a significant emphasis at the next Congress. Water pollution is discussed in chapter 3. The following account owes much to Gifford (1994).

1.3.1 The global warming scenario

Science produces its share of Cassandras whose prophetic forecasts of impending doom to the city are received with some scepticism, especially when the extravagantly alarmist scenarios about the effects of global warming have been subsequently wound back. The uncertainties inherent in the balance of processes producing greenhouse gases and their effects on climate make hard ploughing for the evangelist and recent evidence of the enhanced role of soils under grassland in accumulating refractile carbon provides additional options for the management of global warming.

The evidence is indubitable that global atmospheric concentrations of CO_2 have increased from about 280 ppmv in pre-industrial (1750–1800) times to *c.* 355 ppmv in 1991 (Houghton, Callander & Varney, 1992) and is increasing at the rate of *c.* 1.8 ppmv or 0.5% yr^{-1}. Carbon dioxide is expected to be responsible for more than 50% of the total expected global warming (Minami *et al.*, 1993). Concurrently the other major greenhouse gases CH_4 and N_2O increase at about 1.1 and 0.5% yr^{-1} respectively. Elevated temperature allows more water vapour to be held in the atmosphere and more long-wave radiation will be entrapped unless an increase in water vapour is compensated by increased low cloud cover; whilst clouds may trap *c.* 30 W m^{-2} of infra-red radiation from the earth, they may reflect *c.* 50 W m^{-2} solar radiation back into space and the balance of probability suggests they have a net cooling effect (Roeckner, 1992). Other greenhouse gases with radiative-forcing effect arise from industrial production of chlorofluorocarbons (CFC) and from burning of forest or grassland, which produces small amounts of CO and NO.

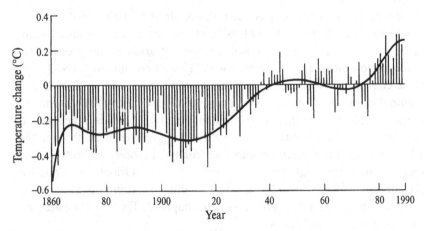

Figure 1.2. Global mean of combined land-air and sea-surface temperatures, for 1861–1989 relative to the average for 1951–80. (From Russell, 1991 and Houghton, Jenkins & Ephraums, 1990.)

Global temperatures have changed in the past, associated with tilting of the earth in relation to the sun, activity of sunspots and volcanoes, and alteration in ocean currents and atmospheric conditions. In the relatively recent past of the Middle Ages there was a warm period between 1100 and 1400 AD with an increase of *c.* 0.2 °C followed by an unexplained Little Ice Age of 1430 to 1850 when temperature decreased as much as −1°C. Since then the earth's atmosphere has moved from the cold late nineteenth century to *c.* 0.3 °C above the average for 1951–80 (Figure 1.2) with abrupt warming in the 1920s and in the past two decades.

There is a strong historic correlation over the past 160 000 years between atmospheric temperature and CO_2 and CH_4 levels (Roeckner, 1992), but this association cannot be viewed in isolation, with other factors held constant, because of several strong feedbacks from increasing levels of greenhouse gases, as further outlined, which impact on temperature, rainfall and the response of man and the biosphere to these changes. This leads to the construction of notoriously complex general circulation models. It is suggested (Houghton, Jenkins & Ephraums, 1990) that a doubling of present CO_2 levels would result in an increase of global mean temperature of 1.9–5.2 °C; a 'Business as Usual' (BaU) scenario predicts an increase of 0.2–0.5 °C per decade during the next century. Several models predict that the greatest warming would occur during winter at high latitudes, and would be greater over continents than over oceans. However, despite the unanimity from thousands of measurement stations concerning increased CO_2 levels, there is great argument as to whether greenhouse-specific temperature or rainfall signals have yet been recorded (Gifford, 1993; Hengeveld & Kertland, 1995). This implies that either feedback processes favouring homeostasis have been underestimated, or that climatic changes will occur as a sequence in future time and not concurrently.

Rainfall is more difficult to predict than temperature, and an additional problem in the lower latitudes is the occurrence of tropical cyclones and variation in the El Niño – Southern Oscillation (ENSO) which generate extreme climatic events (Pittock, 1993). One scenario which brings together five global climate model simulations (Figure 1.3) shows broad agreement on increased rainfall towards the poles from about 50° N and S, but more uncertainty at lower latitudes. Increased rainfall during the northern summer is expected over India and Sahelian Africa, and decreased rainfall is projected for the mid-west of USA, southern Europe, portions of southern Australia, southern Africa and eastern South America. During the southern summer Australia and east Africa would benefit from increased rainfall but eastern China, eastern Mediterranean and some of western Africa, north east Brazil and Mexico might be drier. Daily rainfall intensity would increase, but the return period for heavy rainfall events (i.e. frequency) would decrease (Pittock, 1993). The combined effects of changes in temperature and in rainfall on soil water deficit are still more complex; for example, sensitivity in Mediterranean climates for soil water deficit to summer temperature is expected to be greater than to changed winter rainfall. Climatic change may reduce global production of food grains and exacerbate competition for land between crop and animal production, depending on the agricultural and socio-economic adjustments which occur (Parry & Rosenzweig, 1993).

1.3.2 Carbon flux and grasslands

The uncertainty of climatic prediction is now followed in this discussion by more uncertainty concerning the flows and pools of global carbon, where such large numbers are involved that small changes in particular estimates effect large changes in whether particular components act as C sources or C sinks. The global pools (Goudriaan, 1990, 1992; Minami *et al.*, 1993) of C are estimated as *c.* 39 000 Gt (1 Gigatonne = 10^9 metric tonnes) in the ocean, in excess of 6000 Gt in fossil fuel, *c.* 700 Gt in the atmosphere (of which only 3.5 Gt is methane), and *c.* 1800–2000 Gt in the terrestrial biosphere (of which *c.* 450–600 Gt is in living biomass and the balance is in soil organic matter, OM). The C in the terrestrial biosphere may be subdivided into six ecosystems (Figure 1.4). The high density of soil C under grassland

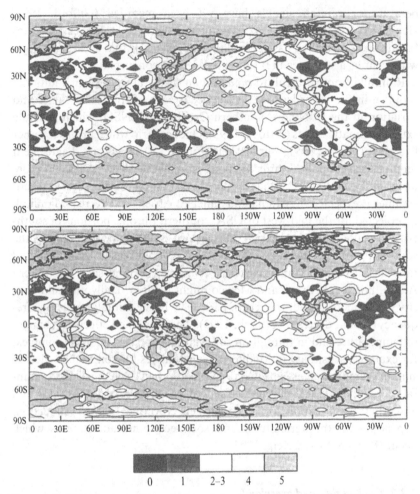

Figure 1.3. Global maps of the number of simulations (out of five) which agree on an increase in rainfall in (a) May–October (northern hemisphere summer) and (b) November–April (southern hemisphere summer) under enhanced greenhouse conditions. Values of zero or one indicate that five or four models agree on a rainfall decrease. (From Pittock, A.B., personal communication 1994.)

and temperate forest, especially of resistant C with a long residence time, is noteworthy, whilst the pool of wood C in forest, especially in relation to the area of tropical forest, constitutes a major C sink. Figure 1.4 also may be used to imply the effects of transfers in land use. The conversion of temperate forest to agricultural land suggests substantial reduction in sequestration of C; a compensating local (but not global) factor may be a concomitant importation of feedstuffs for intensive livestock production

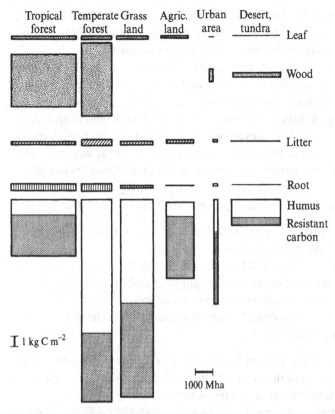

Figure 1.4. Simulated equilibrium distribution of carbon surface densities (heights of columns) for the areas of each vegetation type (widths of columns). (From Goudriaan, 1992.)

on arable land and a consequent soil manurial factor which in the Netherlands, for example, would just counterbalance the losses associated with any translation of forest to arable land (Wolf & Janssen, 1991).

Estimates of the flows between these pools are subject to considerable error. Gifford (1994) summarizes various studies to suggest that global sources of C to the atmosphere constitute 5.8–6.2 Gt from fossil fuel burning and 1.5–6.0 Gt yr^{-1} from tropical deforestation, whilst ocean uptake is 1.3–2.9 Gt yr^{-1} and the atmospheric increase is 3.6–4.0 Gt yr^{-1}, giving a net missing sink of 0.4–4.3 Gt yr^{-1}. Whilst CH_4 molecules may affect global temperature more than CO_2 molecules because of differences in radiative activity and atmospheric residence time, CH_4 is a minor component of the actual C balance sheet, constituting a net source of 0.05 Gt CH_4 yr^{-1} (Minami *et al.*, 1993). This is estimated as arising from 0.45 Gt net sink and sources of 200 Mt CH_4 yr^{-1} from rice fields and natural wetlands, 75 Mt from ruminants, 75 Mt from biomass burning, and 150 Mt from land fill, fossil fuel mining, and other sources such as termites.

Why does the increase in atmospheric CO_2 not match the estimated balance of sources and sinks? An increased estimate of uptake in the biosphere might be balanced by reduced carbon uptake in the ocean, either by reduced exchange between the upper mixed ocean layer and deep sea, or between ocean surface and atmosphere (Goudriaan, 1992); Tans, Fung & Takahashi (1990) argue that no satisfactory theory can be advanced why the ocean might be regarded as an increased C sink and that the biosphere must be accorded a greater role as a sink than previously thought. In assessing this question the fundamental processes to be considered are:

  (i)   net primary productivity (NPP), expressed as the balance between
        photosynthesis and respiration;
 (ii)   the distribution of assimilate to the various plant organs;
(iii)   the senescence of plant organs to form litter and dead wood, their
        consumption by organisms, or their loss to fire;
 (iv)   the fate of litter or excreta in forming humus and refractile C;
  (v)   the decomposition of soil C.

Consideration of the current levels of these processes then needs to be followed by attention to the expected effects of augmented CO_2 and consequent changes in temperature and moisture availability.

The global terrestrial C pool is cycling at *c.* 100–120 Gt C yr^{-1} of gross photosynthesis balanced by varying estimates of autotrophic and heterotrophic respiration (Minami *et al.*, 1993; Gifford, 1994). Net primary productivity has been estimated to vary from *c.* 0.5 t C ha^{-1} yr^{-1} in cold dry climates to 15 t ha^{-1} yr^{-1} in wet, tropical rainforest, and annual or young vegetation have higher rates of productivity than old forest (Pearcy, 1987). It has been discovered that previous estimates under the auspices of the International Biological Programme of grassland NPP (Lieth, 1978, Jordan, 1981) were grossly low. For tropical grasslands the earlier figure of 8 t dry matter, DM, ha^{-1} yr^{-1} should be contrasted with later estimates of 12.4, 17.4, and 22.4 t DM ha^{-1} yr^{-1} respectively at Nairobi, Kenya, Montecillos, Mexico and Hat Yai, Thailand (Long *et al.*, 1989). The early errors arose mainly from a failure to recognize that: (1) whilst the reflectance ratio of red:near infra-red light is useful in estimating the green biomass of growing crops, the amount of dead material in herbage fluctuates and interferes with the technique (Jones, Long & Roberts, 1992); (2) senescence is continuous in humid tropical grassland and is usually highest when growth is most active (Humphreys, 1966b; McIvor, 1984); and (3) shoot production may represent a much smaller source of C than root or rhizome production (Hirata, Sugimoto & Ueno, 1986; Archer, 1993a). Regrettably there is a paucity of data about the flow-on of these effects to soil C levels.

Variation in the partitioning of assimilate between leaf, inflorescence, stem and root is controlled by ontogenetic, climatic and nutritional factors, as discussed later; it is also under genetic control. For example, the grasses *Panicum maximum* var. *trichoglume* and *Cenchrus ciliaris* produced similar net gains of shoot material of 12.1 and 11.5 t DM ha^{-1} respectively in the 21 months from planting in the subhumid, subtropical environment of Gayndah, Queensland; net gains of root growth in the 0–50 cm layer were 4.7 and 10.1 t DM ha^{-1} respectively, suggesting the importance of measuring root production when estimating NPP and its consequent effects on soil OM (Humphreys & Robinson, 1966).

The plant organs formed in grassland may be consumed by herbivores to produce animal products, dung, or methane emitted from rumen fermentation; alternatively, senescing material enters the litter fraction, is consumed by secondary organisms, or is burnt. Some scientists (Korte, 1993) discount the significance of grassland fires for the carbon balance, on the assumption that most herbage material burnt or unburnt eventually appears mainly as CO_2. On the other hand, deforestation and the burning of wood products is a large immediate source of atmospheric C which would otherwise have been sequestered for long periods. The activity of termites produces CO_2, CH_4 and N_2O.

In almost all continuous cropping systems there is a steady depletion of soil organic C (see for example, Ayanaba, Tuckwell & Jenkinson, 1976; Dalal & Mayer, 1986); under grassland a higher equilibrium content of soil C is maintained which is in balance with gains from NPP and losses from the decomposer industry. Soil C merits recognition as having a proportion as refractile humin C with a half-life of *c.* 2000 yr (Vaughan & Ord, 1985). The equilibrium soil C level is subject to great change if constraints to NPP are mitigated by the introduction of elite plant germplasm, the use of fertilizers or the planting of legumes, which augment soil N. Thus, on the low-fertility coastal heathlands of southern Queensland, soil C increased from 0.56% to 1.99% after eight years of improved grassland (Bryan & Evans, 1973). A more recent example with profound implications for global warming (Fisher *et al.*, 1994) is taken from the neo-tropical savanna of the llanos in Colombia, where deep-rooted grasses and legumes have been planted to replace the native savanna dominated by species such as

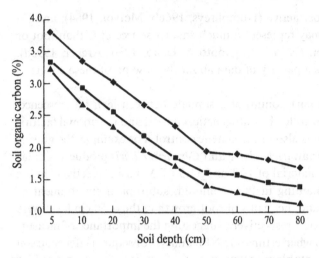

Figure 1.5. Percentage of soil organic C distribution by soil depth at Carimagua, Colombia (▲ native savanna; ■ *Brachiaria humidicola*; ◆ *B. humidicola* with *Arachis pintoi*). (From Fisher, M.J., personal communication, 1994.)

Trachypogon vestitus and *Paspalum pectinatum*. At Carimagua (lat. 5° N, 2240 mm annual rainfall) native savanna was ploughed and fertilized, seeded to *Brachiaria humidicola* with or without the creeping herbage legume *Arachis pintoi*, and grazed by cattle for five years. Soil organic C increased substantially relative to the equilibrium value of the original savanna (Figure 1.5), which contained 197 t C ha^{-1} to 80 cm soil depth. An additional 26 t C ha^{-1} occurred beneath *B. humidicola* alone, and an additional 70 t C ha^{-1} beneath *B. humidicola* with *A. pintoi*, where there was also augmented earthworm activity and a high content of organic P in the earthworm casts. Scientists have usually focused on the OM in the surface soil, but on these deep oxisols 79–89% of the increase occurred below 20 cm depth where it would be inviolable from ploughing or from grassland burning. Similarly, at another site 200 km away at Matazul farm, Puerto Lopez (4° N, 2700 mm annual rainfall), the 0–100 cm layer of an oxisol under native savanna contained 187 t C ha^{-1}, whilst under a fertilized and grazed *Andropogon gayanus – Stylosanthes capitata* pasture an additional 51 t C ha^{-1} had been sequestered 3.5 years after an initial planting with rice. Fisher *et al.* (1994) estimate that there is *c.* 35 million ha of South America planted to *A. gayanus* or *Brachiaria* pastures; although little of this area contains introduced legumes a potentially significant proportion of the missing global C sink may have been located.

1.3.3 Grassland response to CO_2 enrichment

The 'fertilizing' effect of atmospheric CO_2 enrichment on NPP provides a feedback mechanism which potentially reduces the rate at which atmospheric CO_2 levels might increase; the magnitude of this effect on NPP has been variously estimated as +8 to +40% for a doubling of CO_2 concentration (Campbell & Grime, 1993; Gifford, 1994). Short-term experiments with CO_2 enrichment have consistently reported (Campbell & Grime, 1993) increased photosynthesis, shoot production, leaf area, rate of leaf extension, nitrogenase activity and N fixation in legumes, nitrogen use efficiency in growth, water use efficiency and root:shoot ratio; some of these effects are associated with decreased photorespiration and reduced stomatal conductance. There may also be delayed senescence, a possible decrease in dark respiration rate, whilst a problematical decrease in leaf:stem ratio, an increase in C:N ratio, and decreased N% of leaf may reduce the nutritive value of forage; a possible concomitant acceleration of flowering could reduce both nutritive value and the positive response to augmented CO_2.

It is difficult to extrapolate some of the experimental evidence to suggest the same level of longer-term positive responses in the sward situation, which may be of lesser magnitude (Newton *et al.*, 1994 for turves of *Lolium perenne* and *Trifolium repens*). The period of acclimatization and the difference between the responses of plants grown in open chambers or closed growth cabinets affect the conclusions which may be drawn. It should also be recognized that the response to CO_2 enrichment decreases as CO_2 concentration increases, so that the upper levels of projected change will be less successful in increasing NPP. This is illustrated (Gifford, 1994) for three possible scenarios of change (Figure 1.6) in which the standard assumption of 25% increase in NPP when CO_2 doubles from 340 ppmv to 680 ppmv is contrasted with minimal and maximal assumptions; all three curves are based on global NPP of 50 Gt C yr^{-1} at 282 ppmv CO_2. The consequent question is the degree to which increased NPP will result in increased soil C sequestration, as discussed further.

Two of the key issues relate to the response of the plants with the C_4 and the C_3 photosynthetic pathways, and the effects on grass/legume balance. The general assumption is that C_4 plants will be less responsive to augmented CO_2 levels than C_3 plants; this assumption is based mainly on the behaviour of domesticated plants whose physiology has been studied in depth. The reduced oxygenase activity of ribulose-1,5–bisphosphate carboxylase (Rubisco) leads to increased net photosynthesis in C_3 plants (Owensby, 1993), but the absence of measurable photorespiration denies

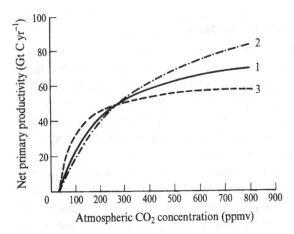

Figure 1.6. Predicted global net primary productivity in response to increased CO_2 concentration (curve 1, standard assumption; curve 2, maximal responses; and curve 3, minimal responses). (From Gifford, 1994.)

this expected response in C_4 plants. However, C_4 grasses also respond to higher levels of CO_2, especially under conditions where the effects of drought are overcome by increased water use efficiency under augmented CO_2 supply, as demonstrated by Nie, He, Kirkham & Kanemasu (1993) for *Andropogon gerardii* in the tall grass prairie of Manhattan, Kansas. Variation in the response of C_3 plants to increased CO_2 is considerable, especially after an initial positive response when canopy elevation occurs. Of grassland plants at Wheldrake Ings in the UK (Watson & Graves, 1993) gains in RGR of *Plantago lanceolata, Anthoxanthum odoratum* and *Trifolium pratense* were not subsequently sustained whilst *Rumex acetosa, Ranunculus acris* and *Holcus lanatus* maintained neutral values and an initial yield advantage due to enhanced CO_2; clearly the competitive relations of species in mixed swards are affected by change in CO_2 level.

Biological N fixation of legumes is directly dependent on the rate of assimilate supply to the nodule if an effective rhizobial symbiosis is in place (for example, Othman, Asher & Wilson, 1988). Consequently, increased CO_2 level which enhances legume photosynthesis increases nitrogenase activity. Crush, Campbell & Evans (1993) report an increase of 55% in nitrogenase activity for *Trifolium repens* when CO_2 level is doubled, and woody legumes respond similarly (Norby, 1987). Additionally, the competitive stress experienced by legumes growing with grasses may be alleviated, especially if the companion grasses are C_4 plants, as illustrated for *T. repens* with *Pennisetum clandestinum* and *Paspalum dilatatum* (Crush & Campbell,

1993). The manipulation of the legume/grass balance has been exacerbated in tropical swards by the differing photosynthetic pathways of the legume and grass components reflected in the greater growth potential of the C_4 grass unless limited by N deficiency; on theoretical grounds the rising CO_2 levels should favour the role of the legume, unless other compensating factors occur; this remains to be investigated.

Combined N in plant material arises from the oxidation of carbohydrate fixed by photosynthesis; if more carbohydrate is produced from the enhanced photosynthesis under augmented CO_2 supply it is logical to expect biological N fixation to track changes in C in the biosphere. Gifford (1994) has referred to the classical Broadbalk wheat experiment at Rothamsted in the UK where wheat has been grown continuously since 1843; total soil organic C and N has been constant for more than 100 years, despite removal of *c.* 30–40 kg N ha^{-1} yr^{-1} in grain, suggesting that non-symbiotic biological N fixation and atmospheric N deposition are maintaining equilibrium.

1.3.4 *Grassland response to changes of temperature, humidity and rainfall*

Increasing CO_2 levels have a secondary effect on plant responses to moisture availability and the expected interactive effects of increased temperature, altered rainfall and nutrient availability need to be considered.

One gloomy prediction is that since plant respiration is sensitive to temperature, whereas at high temperatures gross photosynthetic rate becomes relatively constant, it may be expected that global warming will result in reduced plant assimilation of carbon. These assumptions are partly based on short-term observations. A more hopeful finding is that in fact the ratio of whole plant respiration to gross photosynthesis is remarkably constant over the range 15 °C to 30 °C (Gifford, 1994). He has reported (Figure 1.7) the respiration:photosynthesis (R:P) ratio of seven plant species grown at different temperatures for periods of up to nine months. Despite differences in growth of more than two orders of magnitude the R:P ratio was *c.* 0.4 and independent of temperature; the level of expected change in temperature due to global warming would in these circumstances have no effect on R:P. A further prediction is that, since global temperatures are sub-optimal for the growth of most plant species, increased temperature will therefore result in higher levels of NPP. The greater dominance of warm-season plants and their increased geographical spread is a further prediction which may be unfulfilled if the greater growth response of C_3 temperate grasses to augmented CO_2 level counterbalances the problematical temperature effect.

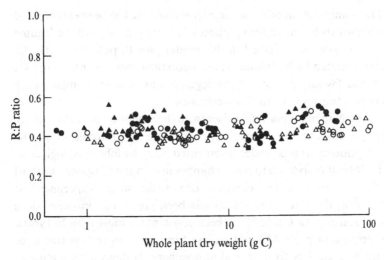

Figure 1.7. The ratio of whole plant respiration to whole plant gross photosynthesis (R:P) over 24 h for seven plant species grown at four temperatures ($\circ$ 15 °C; $\triangle$ 20 °C; $\bullet$ 25 °C; and $\blacktriangle$ 30 °C). (From Gifford, 1994.)

Earlier mention was made of enhanced water use efficiency in response to increased CO_2 level. This operates through reduced stomatal conductance under elevated CO_2 level, which is typically 30–40% in both C_3 and C_4 plants (Woodward, Thompson & McKee, 1991); optimal carbon fixation may occur despite reduced evapotranspiration and reduced water stress in tall prairie plants is accompanied by higher levels of soil moisture (Owensby, 1993). These findings suggest that restriction of plant growth in regions where global warming may result in reduced rainfall may have been overestimated.

These questions of plant response to increased ambient temperature and altered supply of moisture are inextricably linked to the availability and use of nutrients. Enhanced root growth, a more benign rhizosphere with greater density of vesicular arbuscular mycorrhiza (Owensby, 1993) and greater biological N fixation have the potential to increase plant nutrient acquisition; however, the direct measurement of leaf nutrient concentration under increased CO_2 availability indicates protein dilution associated with increased carbon assimilation (Field *et al.*, 1992). Effects on the nutritive value of herbage are discussed later. A different aspect is the predicted effect of a widening of the C:N ratio on the turnover of organic matter.

One group of scientists have assumed that increased temperature coupled with a fixed nutrient pool is likely to release immense amounts of CO_2 to the atmosphere. For example, the Rothamsted model suggested that if the accretion of plant litter remained constant whilst global warming of 0.03 °C yr^{-1} occurred, *c.* 60 Gt C would be released to the atmosphere from the decomposition of soil OM from 1990 to 2050 AD (Jenkinson, Adams & Wild, 1992). Fortunately, this assumption lacks any hard evidence of its occurrence and the balance of theoretical considerations strongly favours the increased sequestration (or at the minimum a maintenance) of soil C under global warming (Gifford, 1992). The first expectation is that an increase in NPP associated with augmented CO_2 level will increase the rate of litter deposition. The second expectation is that the C:N ratio of plant material will adjust to the prevailing conditions for growth and storage. Currently, the ratio is not constant but varies from *c.* 10 in young leaf material to *c.* 500 in refractile humin. Any widening of the C:N ratio in vegetation will reduce the rate of its decomposition and the pool of long-life C in the biosphere will become a still more significant sink.

1.3.5 Management options

The historical association of the level of greenhouse gases and global temperature and the demonstrated increases in the emission of these gases create a social imperative (Pearman, 1988) to ameliorate these effects. Some political activity is oriented to measures (including environmental taxes) which reduce industrial and domestic consumption of fossil fuel, and which ban the manufacture and use of the chlorofluorocarbons CF-11 and CF-12. In this book the focus is on the role of grassland improvement in mitigating global warming, and this requires the dynamic analysis of ecosystems in all of their biological, physical and sociological complexity (Woodmansee & Riebsame, 1993). A basic assumption arising from the review of the situation so far is that the management and maintenance of the capacity of the biosphere to sequester C is a primary objective. This is now discussed in terms of land use, grassland species, fertilizer, fire and stocking policies, the role of legumes and the control of forage quality.

Deforestation

Deforestation was suggested earlier to contribute *c.* 1.5–3 Gt C yr^{-1} to the atmosphere in the 1980s; other estimates are 1.7 Gt C yr^{-1} (Gifford, 1994), and 1.0 ± 0.6 Gt C yr^{-1} (Detwiler & Hall, 1988) for tropical forest clearing. The aerial rainforest phytomass liberates *c.* 140 t C ha^{-1} during burning and decomposition (Serrão, Uhl & Nepstad, 1993); the amount of above-ground residues, which includes refractile charcoal, may be greater after burning than before burning (9.6 after cf. 6.5 t C ha^{-1} before at Manaus,

Brazil; Cerri, Volkoff & Andreaux, 1991). Alternatively, the logging of forest to produce wood products for domestic use results in long-term C residence. The fate of soil C then depends upon subsequent land use. Rapid forest regrowth may restore photosynthetic capacity (Goudriaan, 1992) or annual monocropping systems produce a rapid depletion of carbon, whilst planting of tree crops may be conservative of C.

Some previous assessments have grossly overestimated the net emissions from land clearing. In Australia Burrows (1995) pointed out that the estimates of the National Greenhouse Gas Inventory Committee (1994) failed to take account of the increase in woody plant density and biomass which occurs because of reduced grassland burning and increased grazing pressure, and overstated the extent of actual clearing and of timber burning in savanna woodland. His detailed measurements of tree basal area in grazed woodlands over 24 sites suggest a revision of the estimate downwards of *c.* 130 Mt CO_2 equivalent yr^{-1} for Queensland, or a reduction of *c.* 23% in Australia's net emissions.

The satanic view of pasture development from forest needs to be qualified by the reality of soil C restoration under well managed grassland (Toledo & Formoso, 1993), which at Manaus occurred after eight years (Cerri *et al.*, 1991). Clearing of *Acacia harpophylla* woodland in subhumid, central Queensland and planting to grasses for cattle production led to an average reduction in organic C of 23% in the 0–10 cm layer 12 years after clearing (Graham, Webb & Waring, 1981); however, another study found that soil organic C was rapidly restored to its original level three years after clearing (Lawrence *et al.*, 1993). Many grassland soils have a higher C content than forest soils (Figure 1.4). Emission of N_2O was greater from a forest floor than from grassland in Japan (Minami *et al.*, 1993), but methane uptake by soil was greater under forest than under grassland. There are circumstances under which planted grassland sequesters more C than forest, but the management of grassland determines whether environmental degradation or environmental repair occurs.

Plants species and fertilizer use

Net primary productivity is enhanced by the choice of elite grassland species, which are efficient in their use of water, nutrients and radiation, and reference was also made earlier to differences in the partitioning of assimilate to below-ground organs (Humphreys & Robinson, 1966; Hirata *et al.*, 1986) and in responsiveness to augmented CO_2 level (Nie *et al.*, 1993). Grassland may be classified according to its primary functional type, the more extreme of which are designated as ruderals, which respond to a temporary abundance of resource supply, competitors, which are subject to progressive depletion as resources are exploited, and stress tolerators, which survive under conditions of continuous scarcity (Campbell & Grime, 1993). These functional types vary in their abundance during the established phase and the regenerative phase after disturbance, and may be keyed into various global warming scenarios at a secondary functional level related to phenology to predict vegetation change, as modified by management. For example, it has been suggested (Idso, 1992) that the encroachment of C_3 woody weeds in C_4 grasslands would be favoured by increased CO_2 level, especially if rainfall increases. However, Archer, Schimel & Holland (1995) dismiss any causative effect of the current CO_2 increase on the expansion of woody weeds in the USA. These themes are expanded in chapter 5.

A powerful factor in increasing NPP is the responsible use of fertilizer to remove a constraint to growth. Fertilizer application may also be necessary to maximize the increase in photosynthesis due to augmented CO_2 level; for example, plant yield of *T. repens* was 66% higher in response to CO_2 doubling at good levels of P supply and temperatures of 28 °C day/23 ° C night, whilst plant yield was independent of CO_2 supply at low levels of P supply and at 18 °C day/13 °C, which also interacted apparently to reduce P availability (Carran, 1993). Differences in NPP may be expected to flow on to increase soil C. Thus, Hatch, Jarvis and Reynolds (1991) found that the soil C, which was *c.* 34 t ha^{-1} when pastures were planted, increased to 50 t ha^{-1} 12 years later under a grazed *L. perenne – T. repens* sward, or to 68 t ha^{-1} under a fertilized *L. perenne* sward; the mixed sward was less productive than the fertilized sward and the decomposition rate of residues of legume may have been greater than of grass. The beneficial effects of the tropical legume *A. pintoi* on C sequestration have been illustrated in Figure 1.5 but in that case sole grass pastures were probably limited by N deficiency.

A positive outcome from fertilizer use requires that the associated negative factors are more than counterbalanced. These involve the consumption of fossil fuel and the emissions involved in the manufacture of N fertilizer, and the additional losses of N compounds to the atmosphere when fertilizers containing N are applied to grassland. Thus, the application of farm slurry, diluted, acidified, or untreated, to an organic peat soil in the Netherlands increased N_2O emission (Jarvis, Hatch & Dollard, 1993), and it is suggested that a typical 80 ha dairy farm in the UK has *c.* 540 kg

N_2O-N annual estimated emission. A further pathway of pollution is the drainage of surplus NO_3-N into streams and groundwater, as described in chapter 3. Finally, an increase in NPP from fertilizer use implies increased SR if the additional fodder grown is to be converted to animal products and with concomitant increase in methane emission.

Defoliation

The intensity and frequency of defoliation of grassland, as mediated through herbivory or through fire, modifies the level of NPP (usually negative or neutral), the partitioning of assimilate below ground (negative), the accretion of litter (negative), and the rate of litter disappearance (positive) (Humphreys, 1991).

Burning grassland initially increases the proportion of green leaf in the pasture and pasture quality but reduces the above-ground biomass present (Ash *et al.*, 1982); in south Thailand burning *Eulalia trispicata* grassland reduced NPP, especially through a net loss of roots and rhizomes (Kamnalrut & Evenson, 1992). McKeon *et al.* (1993) predict that increased rainfall of 30% in western Queensland would increase the potential frequency of fires from once in three years to once in two years for spelled pastures. However, the effects of these fires on global warming are problematical. Korte (1993) summarized discussions at the XVII International Grassland Congress as indicating that 'managed burning of grasslands was considered unlikely to increase emissions', but Howden *et al.* (1994) formulated a model whose interpretation depends upon whether the direct anthropogenic effects of increased CO_2, CH_4 and N_2O only are considered, or whether the indirect thermogenic effects of these gases together with CO and NO should also be included. Frequency of burning of grazed natural grasslands in Queensland was respectively related negatively or positively to global warming potential according to which thesis was accepted.

Effects of SR are similarly complex. To the factors enumerated at the beginning of this subsection need to be added not only the effect of animal density on methane emission, but the modification of the level of methane emission from individual animals through effects of SR on pasture quality. Howden *et al.* (1994) produce a convincing model indicating the advantages of conservative SR in reducing greenhouse emissions and increasing output of live weight gain, LWG, per unit of methane emitted. Their study would attribute an average of *c.* 52 kg CH_4 hd^{-1} (head) yr^{-1} for a beast of 300 kg live weight, LW, eating 2.5% LW d^{-1} as DM. Betteridge *et al.* (1993) assess the CH_4 output of New Zealand animals as averaging 76, 58, and 15 kg hd^{-1} yr^{-1} respectively for dairy cattle, beef cattle and sheep.

Legumes

The role of the legume in grassland improvement receives much emphasis in this book and appears in various themes; its significance in mitigating global warming is firstly in relation to the effect of legumes in increasing NPP and secondly in relation to the improvement of diet quality and the reduction of CH_4 emissions from cattle.

Grassland practice may have been too successful (in the context of global warming) in widening the C:N ratio of plant material. For example, the productivity of grass-only pastures decreases with age since planting, since the pool of available N for leaf growth is diminished by the increasing amounts of N immobilized in decomposing litter, unless additional sources of N are supplied (Robbins, Bushell & McKeon, 1989). The primary mineral nutrient limiting grassland production in most grassland situations is N, occasionally overridden by shortage of P or perhaps S (Humphreys, 1994). Biological N fixation (Giller & Wilson, 1991) may be regarded as one engine of increased NPP directed to augmenting soil C sequestration. The increased accretion of soil C under *B. humidicola* pastures in the llanos, to 70 t C ha^{-1} with the companion legume *A. pintoi* or 26 t C ha^{-1} with grass alone (Figure 1.5) should be sufficient evidence of the role of the legume, at least in humid grasslands.

A second significant factor is the maintenance of the nutritive value of forage (Minson, 1990), both where legumes are ingested directly or where they contribute N to ameliorate the expected reduction of leaf:stem ratio and the protein dilution in companion grasses as CO_2 level increases. The benefits to individual animal production reduce methane emissions and increase the ratio of product:methane emission.

Ruminants

The potential usage of the grasslands of the world is limited by the dominance of long-chain structural carbohydrates in the vegetation; ruminants constitute a rare group of animal species which can convert these to products useful to humans. This valuable role is not always recognized in the popular perception of sheep and cattle as environmental contaminators which emit methane to warm the globe and not as residents in a major ecosystem sequestering C. Micro-organisms in aerobic soils constitute the largest biological sink for CH_4, but the sink strength of grasslands is not more than *c.* 8.5% of the total CH_4 consumption. Ruminants are estimated to account for c. 15% of global CH_4 emissions (Minami *et al.* 1993). A dairy farm of 80 ha in the UK might be expected to emit c. 12.2 t CH_4-C yr^{-1} or 152 kg ha^{-1} yr^{-1}, excluding soil deposition of CH_4 (Jarvis *et al.*, 1993). The calculation of methane emission is sensitive to feed quality (Leng, 1993). Rumen organisms ferment ingested forage to produce volatile fatty acids (VFA) and the by products CH_4 and CO_2; the partitioning of digestible energy to VFA and microbial cells and to CH_4 depends on the nutrient balance of the ingested feed, which controls the efficiency of microbial cell synthesis. A second major limitation to production is the level of protected protein which is digested in the intestinal tract rather than the rumen, as discussed in chapter 2.

Methane emission may represent 15–18% of digestible energy when animals are ingesting low quality forage or c. 7–8% if animals are fed urea and mineral supplements. When these values are applied to estimate their relation to product output the enteric CH_4 emitted per unit of LWG or kg milk may be reduced by a factor of four to six. Leng (1993) calculated a schematic relationship between diet quality, expressed as metabolizable energy (ME) kg DM^{-1}, and food conversion efficiency (g LWG MJ^{-1} ME), and derived from this the expected relationship between methane emission and production of milk and meat; the latter relationship, shown in Figure 1.8, implies how attention to improving the nutritive value of grassland or supplementing the diet of the animals grazing it may ameliorate global warming. The impact of this assessment is modified by the weighting ascribed to CH_4 for warming effects, which, relative to CO_2 as unity is 11 for direct effects and 15 for indirect effects according to Houghton *et al.* (1992), or a thermogenic effect of 4 to 6 according to Leng (1993).

1.4 Conclusion

This account has recorded the great uncertainties surrounding the effects of greenhouse gases on climate change. Future research directions (Campbell & Stafford Smith, 1993) are being refined and codified 'to predict the effects of changes in climate, atmospheric composition, and land use on terrestrial ecosystems, including agricultural and production forest systems [and] to determine how these effects lead to feedbacks to the atmosphere and the physical climate system'.

In the agricultural sector the greatest fear arises from the conversion of forest and grassland to annual cropping systems, which decreases both soil C already sequestered and diminishes the future potential of the biosphere to act as a C source, which would delay global warming.

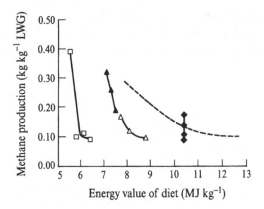

Figure 1.8. Relationship between ME content of diet and CH_4 produced kg^{-1} LWG (straw or ammoniated diet □ Bangladesh; △ Thailand; ▲ tropical pastures with cottonseed meal; ◆ silage with fish protein; ------ UK feeding standards). (From Leng, 1993.)

There are many positive policies available to agriculturalists (Stafford Smith *et al.*, 1995; Stafford Smith, Ojima & Carter, 1996) which can reduce the net rate of greenhouse emissions and the challenge to agricultural scientists is to develop further robust technologies which are both economically feasible and in synchrony with stakeholder objectives. Grazing systems may be viewed as systems of knowledge, and social and economic relationships (McKeon *et al.*, 1993) which determine the extent to which families and institutions can buffer the fluctuations of climate and market imperatives and can accommodate technological innovations which lead to increased NPP and its flow-on to sequestering C. Preeminent amongst these policies are:

(i) The development of cropping systems alternative to annual cropping which incorporate agrosilvopastoral activities or ley pastures

(ii) The replacement of natural grasslands in humid and subhumid areas with planted pastures composed of elite, deep-rooted grasses and legumes, adequately fertilized and grazed at SRs which effect a synchrony between forage allowance and animal requirements in terms which sustain the basic resources of vegetation, soils and animals

(iii) The improvement of diet quality of ruminants grazing on grasslands in order to reduce methane emissions and to increase the efficiency of product output in relation to methane emission.

The challenge to grassland scientists is to devise deft and elegant formulations of modified grassland systems which give due weight to the enhanced efficiency of the long-term use of natural resources for production, the judicious use of external inputs, the amelioration of environmental pollution, and the aspirations of rural (and consumer) communities.

2

The plant genetic base for grassland improvement

2.1 Introduction

The recognition of elite plant germplasm and the constitution of new gene combinations through hybridization have long been fundamental to grassland improvement; simplistically this seeks to optimize the conversion of carbon dioxide, water and minerals to supply digestible herbage acceptable to livestock and to protect environmental resources from degradation.

By the time of the IV International Grassland Congress in 1937 bred varieties from the Welsh Plant Breeding Station at Aberystwyth, such as the late flowering, leafy *Lolium perenne* cv. S23, were well entrenched in agriculture, as were varieties from Sweden, Denmark and other European countries (Osvald, 1937). In Sweden the identification of ecotypes and selection procedures were being applied to legumes such as *Trifolium pratense, T. hybridum* and *T. repens*, especially with respect to winter hardiness. The perennial cry of the seed grower that selection for leafiness led to poor synchrony of flowering and late maturing crops (Evans, 1937) would be reiterated for the next 60 years. Genetic studies with *T. pratense* were sufficiently advanced to describe the location of genes on chromosomes, the inheritance of chlorophyll deficiency, and the occurrence of both self- and cross-incompatibility (Williams, 1937). From the USA Cardon (1937) reported active breeding programmes and studies to define breeding objectives; the impact of bred varieties had yet to emerge.

By the VI International Grassland Congress in 1952 bred varieties were so well established in practice that, in contrast to the IV and V Congresses, there were no papers devoted to the superiority of bred varieties (Williams & Strange, 1952). The brave statement is made that the theoretical and technical bases for the great advances in forage improvement which had appeared in commerce by 1995 were all laid by 1952. Exploration for novel germplasm, selection procedures (especially polycross testing, Frandsen,

1952, and recurrent selection), modes of reproduction (including apomixis, Warmke, 1952), male sterility and compatibility (Brewbaker & Atwood, 1952), quantitative inheritance (Burton, 1952), induced polyploidy (Love, 1952) and interspecific hybridization (Clausen, 1952; Stebbins, 1952) were established. Molecular biology was yet to emerge but had not made any impact on grassland practice by 1995.

Improved tropical grasses were being propagated in 1952, but the commercialization of tropical forage legumes was in its infancy. There were many papers indicating active research in this topic (Strange, 1952 for East Africa; Rowland and Bumpus, 1952 for Zimbabwe; Volio, 1952 for Costa Rica). Christian (1952) referred to the legume based pastures being evaluated under grazing at South Johnstone, Queensland by Graham (1951) and initiated by Schofield (1941). Takahashi (1952) stated that *Leucaena leucocephala* 'is well naturalized and is one of the outstanding tropical forage legumes of Hawaii. It covers many thousands of acres . . .'.

There was discussion of the plant attributes upon which breeding programmes should concentrate (Kramer, 1952) and this has been a continuing thread at all Grassland Congresses (Laidlaw & Reed, 1993).

2.2 Criteria of merit in forage plants
The plethora of plant characters which have received changing attention may be grouped under three interrelated major themes: quality of forage in relation to animal nutrition, attributes of growth, and ecological success in the farm situation.

2.2.1 Forage quality
The major advances over the past 60 years arose from the selection of plants with high animal performance associated with high intake and with better protein delivery systems. The following section owes much to a review by Reid (1994).

Intake and digestibility
Plant introduction and plant breeding have gradually become more oriented to animal performance as the limitations to criteria of plant ecological success alone in promoting farm income have become apparent. Thus, the planting of *Dactylis glomerata*, a grass which was in many respects easier to manage and more persistent than *Lolium perenne* in the UK, decreased as farmers became aware of its inferior quality; Minson,

Table 2.1 *Pasture and animal performance of steers grazing different grass varieties at Aberystwyth, Wales*

	Lolium perenne		Dactylis glomerata	
Attribute	S23	Mascot	Cambria	Conrad
Pasture yield (kg DM ha^{-1})	11690	11240	9580	8680
Pasture intake (kg steer^{-1} d^{-1})	3.62	3.77	3.11	2.91
Grazing days	1230	1130	1010	950
Liveweight gain (kg steer^{-1} d^{-1})	0.80	0.92	0.71	0.77
(kg ha^{-1})	960	1100	790	700

Source: Evans, Munro & Scurlock (1979).

Raymond & Harris (1960) showed that monthly regrowths of *L. perenne* were *c.* five units of digestibility greater than those of *D. glomerata*, and more than five units greater when compared at the same stage of flowering.

These differences may be translated into differences in animal performance, and are also evident in the comparative behaviour of varieties within the same grass species. This is simply illustrated (Table 2.1) by a study of swards at Aberystwyth, Wales (Evans, Munro & Scurlock, 1979) grazed by yearling steers. The yield of pasture, the intake of herbage by individual steers, the rate of LWG and the number of animals carried per unit area were all greater on *L. perenne* than on *D. glomerata* swards. The differences within one species were less than the contrast between species but the LWG ha^{-1} of cv. Mascot, a selection from S23, was *c.* 15% greater than that of the latter; this arose despite marginally lower DM production and mainly because of more efficient use of the nutrients ingested.

Pasture scientists define nutritive value in different ways. Originally, the term related primarily to digestibility and the evaluation of differences in chemical composition. Minson & Wilson (1994) refer to early experiments in which scientists compared feeds at the same level of intake, since it was considered that the interpretation of experiments would become confused if animals ate more of one feed than another. This gave rise to the view that differences in animal performance were often controlled more by differences in voluntary intake than in digestibility (Minson 1971). The product of intake and digestibility is a superior index, but the composition of the nutrients absorbed and their energetic efficiency in maintaining the metabolic functions of the animal need to be considered. These may be assessed in terms of the ratio of fermentation glucogenic energy to total volatile fatty acids, the fermentation protein to energy ratio, and the by-pass nutrient potential (Leng, 1986).

There is a strong but imperfect correlation between digestibility and intake (Crampton, 1957), and since consumption by grazing animals is difficult to measure many scientists have relied on indirect estimates. Intake is controlled firstly by factors associated with the content of nutrients essential for microbial activity in the rumen; thus, intake of forages containing less than *c.* 1.1–1.3% N (Hennessy, 1980) is limited by protein deficiency of rumen organisms which then impacts on the digestion of roughage. The second group of factors relates to the rate of passage of particles from the rumen, and depends on the comminution and packing density of the forage ingested, which is related to many anatomical features (Wilson *et al.*, 1989).

The accessibility of forage to the grazing animal, as determined by its leaf density (Stobbs, 1973a), yield and ease of prehension, is a further component. Voluntary intake is correlated with so many plant characters that it is still not feasible to develop simple predictive equations which can be used in plant improvement. The acceptability of forage to grazing animals remains the primary consideration in plant introduction programmes, since lack of acceptability is the first line of defence in defining weeds which should not be introduced; the breach of this precept has had disastrous consequences in many countries.

The feasibility of selection for high forage quality was realized with the development of laboratory techniques utilizing small plant samples. The old system of proximate analysis was supplemented by better categorization of plant fibre, and the detergent system of Van Soest (1967) was widely adopted. The two stage *in vitro* method of Tilley & Terry (1963) provided better correlation with *in vivo* digestibility than other techniques, but difficulties of standardization led to many laboratories preferring a pepsin–cellulase method (Bughrara & Sleper, 1986) to the rumen fluid – pepsin technique. The application of Near-Infrared Reflectance Spectroscopy (NIRS) promised to constitute a landmark in the rapid assessment of forage quality (Norris *et al.*, 1976). The proportion of vascular bundles and sheaths, the extent to which these are lignified, and the inaccessibility of tissues to micro-organisms are negatively related to digestibility (Wilson, Brown & Windham, 1983).

Hacker (1982) reviewed a number of plant improvement studies and concluded that there was usually variability of ten units of digestibility available, about half of which was genetically controlled. There have been signal genetic advances in herbage quality. Burton, Hart & Lowrey (1967) selected

Coastcross-1 hybrid based on a nylon bag digestibility procedure; this was 12% more digestible than *Cynodon dactylon* cv. Coastal and gave increased LWG under grazing. The trait of lower lignin content, as in brown midrib mutants of *Zea mays* (Barnes *et al.*, 1971), was also introduced into *Sorghum bicolor, S. sudanense*, and *Pennisetum americanum*. These attributes are sometimes linked to unfavourable agronomic traits which require holistic field evaluation.

Protein value

Forage plants differ in their intrinsic N concentration across many environments, as discussed in chapter 3, and this information may contribute to the definition of elite germplasm. Rumen organisms synthesize essential amino acids and this discovery initially led to less focus on protein quality in forages, which was misplaced (Beever, 1993). It was subsequently realized that N digestion in the ruminant was a highly inefficient process (Ulyatt *et al.*, 1975), since much of the protein ingested was degraded as an energy substrate, microbes were unable to capture the released NH_3, and little protein was transferred to the small intestine. Thus, the production of dairy cows grazing high N pastures was increased by supplementing with additional N protected from ruminal digestion (Stobbs, Minson & McLeod, 1977). Differences in the degree of protein protection, as indicated by the degree of protein solubility, are illustrated (Table 2.2) for different tropical herbages (Aii & Stobbs, 1980). Protein solubility was greater for the stem fraction than for leaf, and the range of values was less in the grasses than in the legumes, where the two *Desmodium* spp. tested had very low protein solubility.

One controversial factor associated with protein protection is the content of condensed tannins (Barry & Blaney, 1987), which are also implicated in the absence of bloat. Tannins occurring in the leaves of many shrub legumes form complexes with plant proteins which reduce their degradability in the rumen; this decreases NH_3 production in the rumen and increases the amount of plant protein which leaves the rumen (Norton, 1994). An additional source of protein then becomes available for absorption if the tannin–protein complexes are dissociated in the acid conditions of the abomasum. The alternative worst case scenario is that tannins decrease intake, decrease rumen NH_3 and depress post-ruminal absorption of protein. The protein of some species lacking tannins, such as *Sesbania sesban* and *Albizia lebbeck*, is rapidly degraded in the rumen, whilst species such as *Leucaena leucocephala* contain tannins, provide a high content of by-pass protein, and give superior LWG (Aii & Stobbs, 1980; Quirk, Paton &

Table 2.2 *Protein solubility (%) of leaf and stem of legumes and N-fertilized grasses*

Plant	Leaf	Stem
Legumes		
Desmodium uncinatum	5.3	36.3
D. intortum	7.6	15.9
Aeschynomene indicata	21.0	48.5
Macroptilium atropurpureum	40.8	52.9
Macrotyloma uniflorum	44.7	54.5
Grasses		
Setaria sphacelata cv. Narok	18.6	–
S. sphacelata cv. Kuzungula	19.3	44.2
Pennisetum clandestinum	24.0	–
Digitaria decumbens	24.4	37.9
Panicum maximum	25.7	–
Chloris gayana	29.7	53.6
Panicum coloratum	33.4	43.4
Brachiaria mutica	33.5	37.7

Source: Aii & Stobbs (1980).

Bushell, 1990). Condensed tannins vary in their chemical configuration and the predictive value of condensed tannin analysis for by-pass utilization of protein has yet to be confirmed.

A second factor is the balance of protein and energy required for the metabolic processes of the ruminant (Broderick, 1994). After reviewing recent studies Poppi & McLennan (1995) concluded that losses of protein or incomplete net transfer occur when 210 g crude protein (CP) kg^{-1} digestible OM in forage is exceeded; this represents a degradable CP/available energy relationship for rumen organisms of 13.3, 11.9 or 9.3 g CP/MJ ME respectively for protein degradabilities of 1.0, 0.9 and 0.7. For plants containing protected protein the critical CP concentration would be higher. The opportunity for protein to reach the intestines depends upon an adequate energy supply to the rumen, and low OM digestibility results in wastage of protein, suggesting that improving the net energy value of forages is a desirable if illusory target. Alternatively, the goal of plant selection giving higher energy intake may be more readily attainable.

Other quality indicators

Plant species growing in an identical soil–climate environment may differ radically in the concentrations of minerals in their tissues (Little, 1982;

McDowell *et al.*, 1977, 1983; Minson, 1990), indicating opportunity to recognize germplasm which is elite with respect to its capacity to overcome the mineral deficiencies animals might experience. This depends both on the plant content and bioavailability of the mineral in question, which depends upon the proportion of the mineral absorbed, transported to its site of action and supplied in a physiologically active form (Spears, 1994). For example, Na concentration varies widely in grasses (Griffith & Walters, 1966), Co availability differs between varieties of *Phleum pratense* (Patil & Jones, 1970), and Mg fertilization may or may not overcome Mg deficiency in grasses and legumes (Reid *et al.*, 1978). Mineral deficiencies are often local in occurrence and the appropriate option of selected forage plant, fertilizer application or direct mineral supplementation may be chosen.

Anti-quality factors or secondary metabolites (Rosenthal & Janzen, 1979; Barry & Blaney, 1987) have received increasing attention in plant improvement programmes and their role in herbivore protection has recently been recognized. Thus, the occurrence of endophyte fungi in grasses (Bacon, 1994) may be damaging to animal production, as in *Festuca arundinacea* (Bacon *et al.*, 1977), or may confer resistance to overgrazing and environmental stresses (Bouton *et al.*, 1993a). The minimization of toxic factors such as the amino-acids, mimosine and indospicine, alkaloids, and cyanoglycosides has been sought (Hegarty, 1982), and successful breeding programmes have overcome oestrogenic potency in new varieties of *T. subterraneum* using assays for formononetin (Stern *et al.*, 1983).

2.2.2 *Plant growth*

High rates of plant growth (Osvald, 1937), especially as measured from cutting trials, have long been a primary objective in plant improvement. The biomass produced provides the basic energy supply upon which high SR may be imposed, assists the capacity of plants to dominate their neighbours, and provides the groundcover which facilitates rainfall infiltration and minimizes soil erosion. High seedling vigour contributes to reliable pasture establishment.

These consequences need to be evaluated in the context of animal production, which is the first consideration in most farm situations. There is often a mismatch of DM production and animal response, especially where plant improvement in Europe has been directed to high levels of spring – early summer production for forage conservation under conditions of heavy N fertilizer use. The forage plant giving the lowest DM production may give the highest milk yield, as illustrated in a study by Stobbs (1975) of fodder crops. Table 2.1 also indicates a distinction between DM yield and LWG. Varietal differences in utilized metabolizable energy may be less than varietal differences in DM yield under cutting (Clark, 1993). Superior early growth in height may confer plant dominance but be associated with high levels of lignification. Species compatibility which sustains grass/legume mixtures and provides the biodiversity which provides insurance against shifts in weather conditions, unpredicted disasters, changes in management inputs and emergence of new patterns of disease and pest attack, and greater diversity of diet, may be preferable.

Success in the farm environment needs to be distinguished from ecological success in unmanaged natural environments. Advances in plant improvement may then require the recognition of objectives which run quite counter to the normal adaptive processes and directions of grassland evolution; plant breeders may need to be agricultural rather than Darwinian (Clements, Hayward & Byth, 1983).

Length of growing season

Continuity of forage supply and the consequent minimization of animal stress is desired, but a long growing season may not confer any ecological advantage to the plant if dominance is determined early in the growing season and if plant survival is favoured by cool-season dormancy. Stochastic processes of sward dynamics may emphasize survival rather than production (Breese, 1983).

An extended growing season may reflect greater efficiency of water use (for example, Burton, Prine & Jackson, 1957), which is often associated with exploration of soil moisture at depth. In many environments it reflects greater capacity for leaf growth at low temperatures. This was the foundation of plant breeding activity in the UK arising from plant collecting in southern and western Europe during the period 1939–1959 (Breese & Davies, 1970).

The physiological studies of J.P. Cooper at Aberystwyth of ecotypic response to temperature (Cooper, 1964), daylength (Cooper, 1960) and radiation (Cooper, 1966) were influential. This is illustrated by the behaviour of ten 'climatic races' of *Lolium perenne* collected from Algiers (lat. 37° N) to Lithuania (55° N) and seven accessions of *Dactylis glomerata* collected from Israel (32° N) to Norway (66° N) (Figure 2.1) (Cooper, 1964). This provided a range of 12°C to −6°C for mean temperature of the coldest month. Seedlings were grown in an unheated glasshouse at Aberystwyth. The ratio of the logarithmic rates of leaf expansion in December (mean temperature 6°C) and in May (mean temperature 18°C)

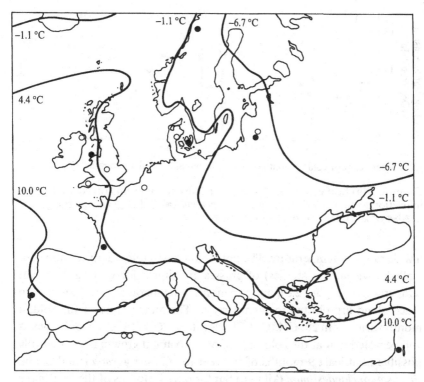

Figure 2.1. Some climatic races of *Lolium perenne* (open circles) and *Dactylis glomerata* (solid circles) in relation to January isotherms. (From Cooper, 1964.)

from successive sowings (Figure 2.2) was positively related to the mean temperature of the coldest month at the site of collection. Rates of leaf expansion were similar under warm conditions, but in cold temperatures the Mediterranean ecotypes had much longer leaves (greater cell expansion?) than those from northern Europe. On the other hand, plant survival after three days at −5 °C was negatively associated with the temperature of origin. The fate of this type of material in subsequently bred varieties depended upon the severity of the winters experienced.

Somewhat similarly Barclay (1960) utilized *T. repens* material from Valencia in southern Spain to improve the winter growth of breeding lines in New Zealand.

Resistance to chilling affects the utility of tropical forage plants grown in the subtropics and in tropical tablelands. West (1970) was the first to identify loss of chloroplast integrity due to starch accumulation in *Digitaria decumbens* as an adverse response to low night temperature. Ivory & Whiteman (1978) differentiated the growth responses of selected grasses to

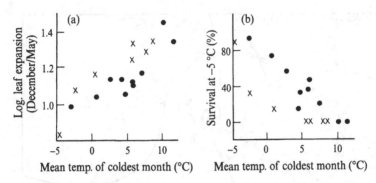

Figure 2.2. (a) Leaf expansion (December relative to May) and (b) cold survival in relation to climate origin of races of *L. perenne* (●) and *D. glomerata* (x), grown at Aberystwyth, UK. (From Cooper, 1964.)

a wide range of temperature. The potential growth rates (derived from controlled environment studies) of grasses are shown for varying day/night temperatures (Figure 2.3), and the monthly temperatures of three sites in Queensland are plotted for comparison. The temperature of the coldest month varies from 20 °C/4 °C at Charleville to 26 °C/19 °C at Cooktown. It will be noted that at the coldest site 25% maximum growth is theoretically possible in at least 9 months of the year for *Chloris gayana* (rhodes) and *Pennisetum clandestinum* (kikuyu) but for only 7 months of the year in the less cold tolerant *Cenchrus ciliaris* (buffel), *Panicum maximum* var. *trichoglume* (green panic) and *P. coloratum* (makarikari). Growth ceased at about 8 °C in *P. clandestinum* and *C. gayana* and at about 12 °C in the other three grasses. These data suggest that some limitation of cold temperature to growth applies even at a coastal lowland tropical site such as Cooktown, at least for some tropical grasses.

Distribution of assimilate

In benign environments plant improvement is oriented to leafiness, since this is associated with forage quality, as mentioned in section 2.2.1, and with growth rate as discussed in chapter 4. Under shaded conditions allocation of assimilate to roots may have negative effects, as discussed with Figure 4.2. However, reliable production of seed is required for commercial adoption of a variety, except in farming communities where vegetative planting is practised. As environments of increasing aridity or frequency of catastrophic events are considered, the choice is between enhanced division of assimilate to a root system which maximizes moisture extraction from the soil and sustains a perennating shoot population, or alternatively

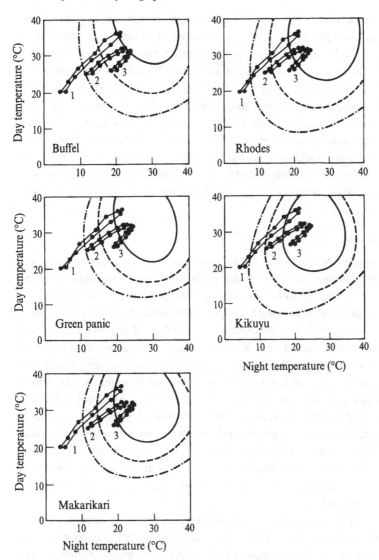

Figure 2.3. Lines of equal growth rate for the levels of 25% (– · –), 50% (– –) and 75% (—) of maximum growth rate for five tropical grasses as related to day and night temperatures, and to mean monthly maximum/minimum temperatures for 1, Charleville (lat. 26° S); 2, Ayr (lat. 20° S); and 3, Cooktown (lat. 15° S) in Queensland. (From Ivory & Whiteman, 1978.)

greater capital investment in seeds for the replacement of short-lived perennial or annual plants. For example, ecotypes of *Cenchrus* spp. from more
arid regions had lower shoot:root ratios and inferior seedling drought survival than those from more mesic environments; the former flowered earlier
and more prolifically (Edye *et al.*, 1964). The role of seed in ruminant diet,
especially legume seeds which assist the animal's N balance in the dry
season, is a further consideration (Playne, McLeod & Dekker, 1972). The
stem component of shrub legumes has a utility for fuel or construction
purposes which is lacking in herbage plants. Selection for the appropriate
balance of assimilate distribution to different plant organs usually arises as
a consequence of attention to other performance indicators in the environment in question.

Net photosynthesis

Most of the genetic advances in crop yield have arisen not from increased
photosynthetic rate but from changed distribution of assimilate to grain,
which does not apply to forage (Austin, 1978). Grassland scientists have
been hopeful about improving herbage performance through application of
growth analysis and the improved understanding of processes associated
with photosynthesis. At the X International Grassland Congress J.P.
Cooper (1966) stressed the significance of variation in canopy characteristics and also reported varietal differences in net assimilation rate which
were positively linked to chlorophyll concentration per unit leaf area.
Subsequently, Cooper & Wilson (1970) referred to substantial, heritable
variation in the maximum photosynthetic rate of single leaves under conditions of light saturation. They predicted that genetic advances in herbage
growth were most likely where water and nutrients were non-limiting (as
confirmed by Pilbeam & Robson, 1992) and where appropriate harvesting
systems were used.

Wilson (1975) analysed the variation in the components of net photosynthesis for *Lolium* and concluded that maintenance respiration may have
been an important determinant of genotypic differences in growth rate.
This led to selection based on low respiration rate, which was expressed in
higher growth rate and increased tiller density. The late-flowering cultivar
Caradoc, which had been bred for reduced respiratory losses, showed high
dry matter yields, especially in late summer (Institute for Grassland and
Animal Production, 1988) but made little impact on European grassland
production; there is a problem of genetic drift in this character in outbreeding herbage plants such as *Lolium perenne*.

An alternative approach is to focus on maximizing net photosynthesis through manipulation of canopy structure to improve light utilization. I. Rhodes (1973, 1975) found that the growth of swards subjected to infrequent defoliation was greater from morphological types with long leaves, erect tillers and rigid leaves. Under frequent defoliation, lines with short leaves and prostrate tillers were better adapted to the maintenance of residual leaf area and efficient light use. He identified significant genotypic variation in leaf length and tiller angle in *Lolium*. There is a seasonal variation in the canopy ideotype for temperate grasses. In spring when flowering promotes stem elongation young leaves are situated near the top of the canopy where these can develop in full sunlight. In summer and autumn newly expanded leaves are situated at the base of the vegetative canopy and have a lower photosynthetic capacity, since they are developed in a low light environment (Woledge & Leafe, 1976; Parsons & Robson, 1982), especially if long narrow leaves shade those below them. Breeding activity has therefore to be 'judged in the context of specific management procedures which can be related to realistic farm systems' (Breese, 1983).

The differences in photosynthetic rate discussed so far pale into relative insignificance when compared with the contrast between the C_3 and C_4 pathways. In C_3 temperate grasses and temperate and tropical legumes the incorporation of atmospheric CO_2 into photosynthate follows the Calvin–Benson cycle (Calvin, 1962), in which the 3-carbon acid phosphoglyceric acid is the first product; the carbon dioxide acceptor molecule is ribulose diphosphate, with the companion enzyme, RUDP carboxylose. C_4 plants have an additional primary cycle in which the first products of photosynthesis are the 4-carbon acids malic and/or aspartic acids, as discovered by M.D. Hatch and C.R. Slack (1966). The CO_2 acceptor molecule is phosphoenolpyruvate, which is more reactive with CO_2 than the RUDP acceptor system. There is no photorespiration, and C_4 plants respond positively to much higher levels of illuminance than do C_3 plants (Ludlow & Wilson, 1971a). The 'Kranz' anatomy leads to efficient use of assimilate and effective water transport. The bundle sheaths may be categorized according to whether the species belongs to the C_3/C_4, PCK, NADP-ME or NAD-ME C_4 groups (Wilson & Hattersley, 1983).

The C_4 grasses often grow faster than C_3 plants by a factor of 1.5–2, and have higher temperature optima for growth and lower transpiration ratios, but their lower digestibility (Wilson & Minson, 1980) illustrates the dichotomy between the objective of increased DM production and acceptable levels of animal performance.

N fixation in legumes

This primary selection attribute for forage legumes is closely associated with growth rate, as discussed in chapters 3 and 4, and there is also considerable scope for breeding compatible host and bacterial genotypes (Mytton & Skøt, 1993) which fix more N under field conditions in competition with indigenous rhizobial strains and in the presence of relatively high levels of soil nitrogen.

2.2.3 Ecological success in the farm situation

Farmers value forage plants which are persistent under the management system imposed on the farm and the variation in environmental conditions which occur there. The expense of replanting is avoided if plants exhibit persistence of yield and resistance to the invasion of unwanted plants; perhaps incursion of the favoured species into unplanted areas is desired.

Ecological success may be conveniently considered in terms of the capacity of plants to interfere with the availability of environmental growth factors to their neighbours, and/or greater resistance to environmental stresses, which may be categorized in climatic, edaphic and biotic terms (Humphreys, 1981). Superior performance is expressed in superior survival of reproduction and plant replacement (Harper, 1977, 1978). The value accorded ecological success is modified by the extent to which this depreciates pasture quality and animal performance, and compatibility in desired mixtures.

Compatibility and competition

Plant biodiversity is preferred to monospecific swards by many farmers, for reasons mentioned earlier. Coexistence is sought especially for grass/legume mixtures. A special problem occurs in the tropics where C_4 grasses tend to dominate C_3 legumes, unless N deficiency limits grass growth or the legume is less grazed; both situations have undesirable features. The incompatibility is overcome in shrub legume/grass associations, where the added stature of the shrub precludes competition for light (Gutteridge & Shelton, 1994).

Compatibility arises from spatial and/or temporal components (Rhodes & Webb, 1993). Species may have complementary growth rhythms which minimize the degree of competitive stress. Alternatively, they may explore the soil mass to a differing degree in different layers, or their shoot morphology may enable radiation to be shared. The relative independence of the legume with respect to soil N may compensate for other deficiencies.

Table 2.3 *Total seasonal dry matter yield of grass/clover mixtures at Aberystwyth, Wales*

	Yield (g m^{-2})		
Associate grass	Clover	Grass	Mixture
S23	510	234	744
BA9462	642	296	938
Mean of 5 coexisting populations	709	320	1029

Source: Evans *et al.* (1985).

The dominance of grass in association with clover, which has limited the continued acceptance of clover in the UK, has been partly attacked by the identification and development of more compatible varieties. Evans *et al.* (1985) noted that *L. perenne/T. repens* mixtures based on components which had previously coexisted (coadapted?) had a higher proportion of clover than in other mixtures. The yields of five grass/clover mixtures which had previously coexisted for a long period (Table 2.3) were compared with the yield of mixtures containing the same clover components grown with the variety *L. perenne* S23 and with BA9462 (Evans *et al.*, 1985). In the second year after sowing the swards were clover dominant; the clovers grew better with their coexisting grasses than with other grasses, and grass yield was improved, suggesting a greater transfer of N from *T. repens* to grass. This finding that coevolution contributes to compatibility (Collins & Rhodes, 1989) is being utilized in breeding programmes.

The adaptation of herbage plants to shaded conditions is crucial for their coexistence in agroforestry and has received attention in the identification of pasture grasses and legumes for the understorey of tropical plantations (Wong, Sharudin & Rahim, 1985; Wong, Rahim & Sharudin, 1985). Most improved pasture species have been selected from sites in full sunlight, but there are 'shade' species (Wong, 1991) which exhibit reduced respiration rate in shade, a greater diversion of assimilate to shoot rather than root, and increased specific leaf area which enhances light interception and modifies growth responses (Wilson & Ludlow, 1991).

Resistance to climatic stresses

Resistance to stress has been a fruitful area of interest, and is simplistically considered in terms of plant escape, plant avoidance of stress, and plant tolerance of the stress actually applied (Ludlow, 1989).

Plant adaptation to drought has been recognized in fast-growing annuals which complete their life-cycle before soil moisture is exhausted. New mechanisms of drought avoidance, such as paraheliotropic movement which decreases leaf temperature (Ludlow & Björkman, 1984) and early closure of stomata (Sheriff & Kaye, 1977), were observed and herbage legumes with contrasting tolerance of desiccation and capacity for osmotic adjustment were identified (Ludlow, 1980). Carbon isotope discrimination now offers a technique for screening the efficiency of water use in temperate grasses (Johnson *et al.*, 1990).

Adaptation to freezing is crucial for plant survival in higher latitudes and is a continued preoccupation of plant breeders (Simonsen, 1985; Kunelius *et al.*, 1993). Sprague (1952) and Smith (1950) linked freezing resistance in temperate legumes such as *Medicago sativa, T. repens* and *T. pratense* to the content of non-structural carbohydrate in the roots and crown. For some grasses this appeared to be associated with fructan accumulation, as in *Phleum pratense* (Klebesadel & Helm, 1986; Suzuki, 1993) and *Dactylis glomerata* (Kobayashi *et al.*, 1993), but alcohol-soluble carbohydrate content was not correlated with freezing resistance in *Setaria sphacelata* (Hacker, Forde & Gow, 1974). Freezing injury is associated with intrinsic characteristics at the cellular level (Burke *et al.*, 1976) but may also be associated with plant habit, such as the height of the lowest bud (Clements & Ludlow, 1977).

The dangers inherent in breeding for growth at cool temperatures in areas with harsh winters are illustrated in Figure 2.2. A successful combination of freezing resistance and high rates of leaf growth in cool spring conditions appears to have been found in Swiss material of *T. repens,* which has led to the release of the variety AberCrest (Rhodes & Webb, 1993). A series of studies identified winter stolon survival as the key factor determining the persistence of yield in *T. repens* in the UK (Collins, Glendining & Rhodes, 1991). Stolon characteristics may be expressed as length per unit area, which is relatively easy to measure, the bud density per unit stolon length, and stolon weight per unit length, which was consistently related to annual clover yield. The stolon survival in winter (Table 2.4) was poorer in the widely sown variety Huia than in AberCrest, which additionally displayed superior leaf growth at low temperature (Rhodes & Webb, 1993). The trigger for spring leaf growth is an interesting phenomenon of increasing daylength (Rhodes *et al.*, 1993).

Resistance to edaphic and physiographic stresses

Adaptation of plants to acid soils and their associated toxicities, and the problems of micronutrient deficiencies exacerbated by liming soils, are dis-

Table 2.4 *Indicators of winter damage and low temperature growth of*
T. repens *at Aberystwyth, Wales*

	Clover variety	
Character	Huia	AberCrest
Stolon length per unit ground area (m m^{-2}):		
November	158	144
March	57	103
Leaf yield per unit stolon length at 6 °C (mg mm^{-1})	2.0	3.5

Source: Rhodes & Webb (1993).

cussed in chapter 3. Resistance to other toxicities such as salinity or excess
heavy metals are sought. There is also a basic dichotomy of approach
regarding the emphasis given to successful plant adaptation to low fertility
conditions or to superior growth response to improved mineral nutrition,
which is also mentioned in chapter 3.

Adaptation of plants to soil with poor internal drainage is often linked
to resistance to flooding, since anoxia is the dominant constraint. Some
examples are given in Humphreys (1981).

Resistance to biotic stresses

Resistance to grazing features in chapters 4, 5 and 6. Plant improvement in
many countries is focused on resistance to diseases (Lenné & Trutman,
1994) and pests. On the one hand the relatively low prices for animal products predicate the cheaper option of plant resistance and on the other hand
the presence of chemical residues in food products affects consumer and
institutional acceptance. These approaches are linked to the management
of natural predators which alleviate the problem.

Three illustrations are given. The success of *T. repens* in the UK is constrained by a wide suite of pests and diseases (Hopkins, Davies & Doyle,
1994). In one survey (Lewis & Thomas, 1991) leaf damage by slugs varied
from 23 to 67% and by *Sitona* weevils it ranged from 3 to 62%. Aphids, leaf
and seed weevils (*Hypera* and *Apion* spp.), gall midges (*Dasineura* spp.) and
leather jackets (*Tipula* spp.) are also prevalent. The main disease is clover rot
(*Sclerotinium trifoliorum*), to which varietal resistance (cvs. Alice and Siwan)
has been attained through breeding (Rhodes & Webb, 1993; Institute of
Grassland and Environmental Research, 1994). Varietal differences in susceptibility to slug damage and to nematodes have also been identified.

Some key events in the use of *Stylosanthes* spp. as a herbage legume illustrate the evolution of approaches to the development of disease resistance. *S. humilis* was a chance introduction from South America to Australia, where it was first recognized *c.* 1903 (Humphreys, 1967). This grazing resistant, low-fertility-demanding plant was promoted by scientists and farmers; local ecotypes which differed in their length of growing season were identified (Cameron *et al.*, 1977) and by 1975 it had been planted or become naturalized on at least 0.5 M ha in Australia. The arrival of the disease *Colletotrichum gloeosporioides* (anthracnose) (Irwin & Cameron, 1978) devastated *S. humilis*, and cultivars of *S. guianensis* were also variously affected by the different pathogenic races which evolved. From the plant introduction programme two replacement cultivars emerged: *S. hamata* cv. Verano, with field resistance to the disease, and *S. scabra* cv. Seca with major gene resistance. By 1993 the former occupied *c.* 0.5 M ha and the latter *c.* 0.3 M ha of Australia (Cameron *et al.*, 1993a).

There were then several strategies adopted in breeding programmes. Because of the inherent risks of relying on cultivars developed from a narrow genetic base in dealing with such a rapidly evolving organism, collections were made from the host centre of origin where resistances may have coevolved. The second approach was that of the multiline, and CIAT (Centro Internacional Agricultura Tropical) released *S. capitata* cv. Capica as a mixture of five lines selected for anthracnose resistance and diversity of geographic origin (Toledo, Giraldo & Spain, 1987). Somewhat similarly in Australia *S. scabra* cv. Siran was released as a composite of three lines developed by breeding and selection from four different sources of resistance (Cameron *et al.*, 1993b).

An alternative approach is based on the use of partial or quantitative resistance, which has shown stability and attributes which are rate-limiting for disease development. Variation in the course of disease development (Figure 2.4) at Southedge, Queensland (lat. 17° S) is strikingly evident in lines of *S. scabra* and in *S. hamata* cv. Verano (Chakraborty *et al.*, 1990). Crossing and recurrent selection under conditions of artificial exposure to the pathogen show continued positive responses. Concurrently, at a more sophisticated level research proceeds on the development and use of molecular markers, the identification of pathogenicity genes in the organism, and the subsequent transformation of novel disease resistance genes into *Stylosanthes* (Chakraborty, Cameron & Lupton, 1996).

The third illustration is the response to the attack of the psyllid *Heteropsylla cubana* as it moved from America to Asia and Australia (Bray & Sands, 1987) to Africa, devastating *Leucaena leucocephala*. Several years

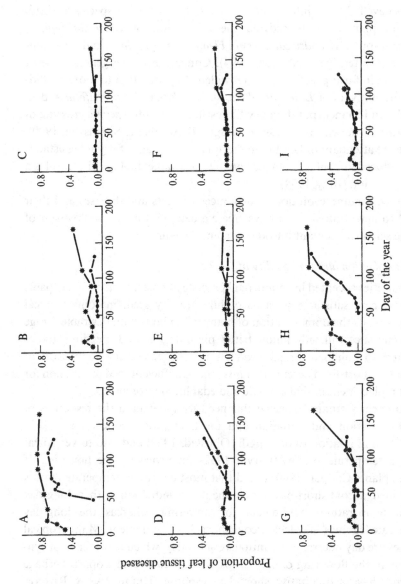

Figure 2.4. Anthracnose development on various accessions of *S. scabra* (susceptible control cv. Fitzroy (A), resistant control 93116 (C) and others (B, and D–H)) and on *S. hamata* cv. Verano (I) during 1988 (●) and 1989 (*) at Southedge, Queensland. (From Chakraborty *et al.*, 1990.)

after its arrival a new homeostasis has developed between the host and the pest, which is less damaging than the initial ravages of the epizootic, and the outbreaks have only been severe in humid, warm conditions. These events have led to an interest in increasing the number of species of shrub legumes planted, and studying the adaptation of *Gliricidia sepium, Sesbania* spp., *Calliandra calothyrsus, Flemingia macrophylla* and other candidates (Gutteridge & Shelton, 1994). Concurrently, the genetic base of *Leucaena* in farm practice has been widened by attention to more psyllid-tolerant cultivars of *L. leucocephala,* such as K636, *L. diversifolia* and *L. pallida,* and to incorporating psyllid resistance by interspecific crossing of *L. leucocephala* with other *Leucaena* spp. (Brewbaker & Sorensson, 1990). There are attendant difficulties in the apparent linkage of psyllid resistance with reduced animal acceptance and nutritive value; biological control is a further option (Bray, 1994).

The continuing evolution of new races of pests and diseases and their travel to new centres will always create a demand for the employment of forage scientists in plant introduction and breeding.

Seed production and seed quality of forages

The varieties accepted in commercial practice almost invariably are capable of producing satisfactory yields of high quality seed; in some tropical environments the domestication of many wild plants with desirable forage characteristics will only follow better physiological and breeding studies directed to enhancing seed production. The ecological adaptation of annual or short-lived perennial plants reflects sufficient seed production for plant replacement in that climatic and edaphic environment.

Many early studies reflected this need to understand the responses of floral induction and initiation and the later processes leading to seed formation in relation to daylength (Olmsted, 1952) and also to vernalization in temperate species (Purvis, 1952). In reviewing the flowering of forage plants Cooper (1960) noted that most cultivated temperate grasses combined a cold/short-day requirement for induction with a long-day warm temperature requirement for flowering, whereas the long-day temperate legumes may lack a cold/short-day requirement. Many tropical grasses are day neutral or quantitative short day, which causes lack of synchrony in the flowering of seed crops. Several cultivated tropical herbage legumes have a qualitative short-day response (Humphreys & Riveros, 1986); the critical daylength for flowering is extended by cool temperatures (Ison & Humphreys, 1984) and a long–short-day requirement occurs in some varieties of *S. guianensis* (Trongkongsin & Humphreys, 1988).

Even in the tropics the sequence of flowering of varieties reflects a tight control through variation in daylength. Temperature controls the rate of processes resulting in seed formation (Ryle, 1966). Mild moisture stress facilitates flowering in *Macroptilium atropurpureum* (Kowithayakorn & Humphreys, 1987).

Realization of the yield potential of forage plants (Hill & Loch, 1993; Lorenzetti, 1993) depends first upon the choice of site for seed production, which can reflect the climatic requirements for the particular variety (Hopkinson & Reid, 1979; Ferguson *et al.*, 1983). Selection for high seed yield within outbreeding populations of grasses is feasible (Griffiths, Lewis & Bean, 1980). Increased density of inflorescences and their associated stems may depreciate forage quality, and the focus of plant improvement is on the efficiency of the reproductive system (Bean, 1972), as expressed, for example, in enhanced ovule fertility in *T. repens* (Pasumarty *et al.*, 1993) and reduced seed shattering in *Festuca arundinacea* (Falcinelli, 1993). Selection for high individual seed weight is regarded as justified only when this factor impacts on seed quality (Lorenzetti, 1993).

Herbage seed quality derives from genetic integrity, the purity of the sample and from viability, vigour and longevity (Roberts, 1972). The mode of reproduction conditions the requirements for spatial isolation of seed crops (Griffiths, 1956). Physiological studies have focused on the effects of harvest procedures and storage environment on the maintenance of seed viability; grassland ecologists have recognized the balance of processes associated with the seed coat or the condition of the embryo which control seed dormancy (Taylorson & Hendricks, 1977) as significant in the maintenance of reserves in the soil of seed which are capable of responding to favourable conditions for establishment. Climatic conditions during seed formation influence the development of seed dormancy (Argel & Humphreys, 1983).

Criteria of merit are recognized in the creation of recommended lists of varieties, many of which are protected by plant variety rights and seed certification programmes (Hides & Desroches, 1989). This protection has increased investment in plant breeding research. It is hoped that the motivation for high DM yield performance in short-term cutting trials does not obscure the perspectives of high animal production and environmental protection which are inherent in the ethos of responsible grassland improvement.

2.3 Conservation of germplasm and plant introduction

2.3.1 The rationale for conservation of germplasm

Scientific and public interest groups share common ground in seeking conservation of the world's germplasm, but the motivation for exploration of particular habitats, the criteria for collection of material and its documentation, the management of information concerning the collections, their preservation, and their evaluation reflect different thrusts.

Genetic erosion

The preservation of our global genetic heritage is a primary need and has attracted especial attention as plant species have been lost through changed land use. The plant and animal genetic base which has evolved to the present reflects a store of environmental fitness which needs to be guarded against further depletion. These threats arise as forest and woodland are converted to other purposes, as grasslands become crop land and as urbanization completely destroys native habitat. They may also arise as land use changes within a particular category. The overgrazing of natural grasslands leads to loss of species; the wider application of weedicides in Mediterranean crop land threatens the extinction of valuable native herbage legumes which volunteer in wheat and barley fields (Reid, 1993). In Brazil it has already not been feasible to recover *Arachis pintoi* and *Centrosema acutifolium* germplasm from the well-documented, original collection sites of their botanical type specimens (Schultze-Kraft, Williams & Keoghan, 1993).

The preservation of relict areas and of wilderness reserves for recreation is one response to genetic erosion, as mentioned in chapter 1, and gains political will as an increasingly urban population seeks recreation in open spaces. The alternative is plant collection and storage, although genetic erosion continues if seeds are stored unsuitably (Roberts, 1972).

Biodiversity

The preservation of biodiversity in communities is a slightly different thrust. It also involves the stewardship of genetic resources, but has the added impetus that sustainability is seen to reside in diversity (Hadley, 1993). The maintenance of system structure and the efficiency of resource use are considered to rely on biodiversity for the functioning of the ecosystem, especially under conditions of fluctuating weather and of the occurrence of catastrophic events.

Geographical barriers to plant evolution

The adherents of the view that the primitive condition is best, since the natural, original grasslands of any locality must represent an evolutionary peak, need to take account not only of recent changes in botanical composition but of the geographical barriers which have interrupted the progress of global plant evolution. This is especially pertinent for grasses and legumes, which are not readily transported by natural means across mountain barriers, oceans or regions of differing climate. Geohistory is also involved; the separation of the continental plates isolated plants, and changed the climatic conditions under which they evolved. These barriers are reflected in the endemic character of many subtropical species. For example, the valuable grass *Paspalum dilatatum* apparently originated in south-east Brazil, northern Argentina and Uruguay (Bogdan, 1977), but was not present in similar latitudes of northern America until introduced there in the nineteenth century by human agency.

The success of plant introduction in raising grassland production and protecting the environment through the provision of plants more resistant to grazing, drought, and unfavourable soil conditions has become accepted by farming communities in diverse agricultural situations and its extensive documentation (for instance, Mathison, 1983; Toledo, 1985; Walker & Weston, 1990; Teitzel, 1992) bears study by some environmental lobbies seeking to preserve the primitive condition as a general land use precept.

Altered land use

The greatest benefits from plant introduction arise in disturbed habitats, and the greater the change the greater is the need for and response from introduced species (Williams, 1983). Diminutive shade tolerant plants of woodland may be less productive than introduced plants sown when trees are removed. The naturally occurring *Paspalidium* spp. in Queensland do not compete successfully with the more vigorous *Cenchrus ciliaris, Panicum maximum* var. *trichoglume* and *Sorghum almum* when *Acacia harpophylla* woodland is removed (Johnson, 1964). Plants will be sown in areas whose soil fertility may be increased by cultivation, fertilizing or the incorporation of legumes.

Plant improvement programmes

The breakdown of introduced varieties when their natural predators are also introduced to the new environment was mentioned earlier in this chapter. It is difficult to predict the new threats which may emerge for plants currently successful, and a widely based germplasm bank provides insurance. Many plant improvement programmes have defined objectives which are met by targeting collecting environments which provide plants meeting the particular criteria of merit sought.

2.3.2 The development of collections

Early activity

Much of the early accumulation of germplasm in banks depended on rather random postal exchange, but plant exploration was well established in scientific culture in the nineteenth century. Smith & Quesenberry (1993) cite a reference indicating that in 1897 the USA had initiated a systematic collection of forage legume germplasm in the centres of origin of the species of significance to local agriculture. Scientific collaboration in collection developed between institutions and FAO held its first international discussions on plant genetic resources in 1947 (Reid, 1993). Reference has been made to the enrichment of UK, European and New Zealand forage plant improvement by early collecting of Mediterranean germplasm. William Hartley travelled from Australia to Uruguay, Paraguay, Argentina and Brazil in conjunction with the USDA in 1947–1948; from this extensive collection *Stylosanthes guianensis* var. *intermedia* cv. Oxley, *Aeschynomene falcata* cv. Bargoo and *Paspalum plicatulum* cv. Hartley emerged into tropical grassland practice (Burt & Williams, 1975). This was followed by other tropical collecting missions in Africa and Central and South America by J.F. Miles, W.W. Bryan, R.J. Williams and D.O. Norris (for rhizobia) which led to further releases, and systematic collecting, storage and evaluation became an integral function of grassland research.

The present collections

Reid (1993) refers to 189 collections held globally in 73 countries, which embrace 384 000 accessions from 3644 separate species. These comprise 189 000 grass accessions, 167 000 legume entries, 8000 browse and all of 19 000 unknown entries; the last reflects the lack of knowledge of intrinsic value about much of the stored material. The New Zealand Germplasm Centre, the Vavilov Institute, CIAT and the CSIRO (Commonwealth Scientific and Industrial Research Organization) Division of Tropical Crops and Pastures, Brisbane have the largest collections, and eight other banks each hold more than 10 000 accessions. The International Centre for

Agricultural Research in the Dry Areas (ICARDA), Aleppo, Syria special-izes in Mediterranean legumes and comprises *c.* 16 500 accessions of 19 genera (Schultze-Kraft *et al.*, 1993). The main genera are *Astragalus, Lathyrus, Medicago, Trifolium, Trigonella* and *Vicia*.

There are many collaborative networks designed to share both germ-plasm and the learning process. A project based at ICARDA augmented the IBPGR Mediterranean Forages Database to 34 000 records of 720 forage taxa. The European Cooperative Program/Genetic Resources had met thrice by 1989 to coordinate documentation of resources and to develop a number of passport descriptors. The AFRC Institute of Grassland and Environmental Research at Aberystwyth maintains an active bank of *c.* 10 000 accessions of more than 450 species of temperate forage grasses and legumes, mainly of *L. perenne, Dactylis, Festuca, T. repens* and its associated *Rhizobium trifolii* (Sackville Hamilton *et al.*, 1993).

The temperate forage germplasm collections, although larger than the tropical collections, reveal a narrow focus, which reflects their emphasis on plant breeding programmes (Schultze-Kraft *et al.*, 1993). The cultivar list for OECD countries, New Zealand and Australia indicates that 94% of temperate grass cultivars are from a limited number of species in eight genera (*Lolium, Festuca, Poa, Dactylis, Phleum, Bromus, Agropyron* and *Phalaris*) and 94% of temperate legume cultivars are from *Trifolium, Medicago, Lotus* and *Vicia*. Schultze-Kraft *et al.* (1993) regard the major-ity of species as adapted to relatively high-input grassland systems in which the environment is ameliorated to favour the planted species.

The tropical forage collections reflect the shorter research history of these regions and the wider interest in what are mainly still 'wild' plants. Schultze-Kraft *et al.* (1993) estimate these as *c.* 600 grass species (6000–7000 accessions) from 100–120 genera, and *c.* 1500 legume species (26 000–28 000 accessions) from 150–200 genera. The 20 largest genera (in terms of accessions) held by the three main institutions (Tables 2.5, 2.6) show 15 grass and 10 legume genera in common, arising from both research interest in these genera and their wide natural distribution and availability. The CIAT collection reflects its commitment to finding adapted legumes for acid, infertile soils whilst the ILCA collection indicates special involve-ment in shrub legumes and in tropical highland species (Hanson & Lazier, 1989). Significant collections also exist at the Nitrogen-Fixing Tree Association, Hawaii, and at the University of Florida's centre at Fort Pierce.

Table 2.5 *Tropical grass germplasm: the 20 largest collections in each of the major genebanks, in terms of numbers of species (spp.) and accessions (acc.)*

Genera	CSIRO		CIAT		ILCA	
	spp.	acc.	spp.	acc.	spp.	acc.
Andropogon	21	106	3	100	6	45
Anthephora	8	160	–		–	
Aristida	–[a]		–		6	45
Axonopus	–		3	16	–	
Bothriochloa	13	214	3	20	–	
Brachiaria	23	169	27	687	27	658
Cenchrus	11	536	2	54	4	114
Chloris	23	193	4	55	7	114
Cymbopogon	3	38	–		–	
Cynodon	7	70	2	15	3	107
Dichanthium	4	144	–		–	
Digitaria	46	425	8	29	13	53
Echinochloa	10	63	5	15	5	54
Eragrostis	17	140	7	55	13	59
Festuca	–		–		7	38
Heteropogon	–		2	27	–	
Hordeum	–		–		1	69
Hyparrhenia	9	59	13	71	8	38
Lolium	–		–		5	75
Melinis	–		1	10	–	
Panicum	58	632	10	536	16	197
Paspalum	50	339	14	114	9	63
Pennisetum	20	334	9	53	18	210
Sehima	4	40	–		–	
Setaria	35	268	5	47	8	65
Sorghum	20	93	5	11	5	68
Sporobolus	–		2	12	9	28
Urochloa	7	217	4	24	7	33
Total	(20)[b] 389	4240	(20) 129	1951	(20) 177	2133
Other genera	(96) 127	564	(27) 31	50	(58) 98	361
Grand total	(116) 516	4804	(47) 160	2001	(78) 275	2494
% 'Total' of 'Grand total'	75	88	80	97	64	85

Notes:
[a] included in 'Other genera';
[b] number of genera.
Source: Schultze-Kraft *et al.* (1993).

Table 2.6 *Tropical legume germplasm: the 20 largest collections in each of the major genebanks*

Genera	CSIRO		CIAT		ILCA	
	spp.	acc.	spp.	acc.	spp.	acc.
Acacia	—[a]		—		67	178
Aeschynomene	31	457	31	983	—	
Alysicarpus	14	353	9	256	9	175
Cajanus	—		—		1	157
Calopogonium	—		4	526	—	
Canavalia	—		6	231	—	
Cassia (sens. lat.)	72	292	14	225	—	
Centrosema	32	1231	33	2376	12	326
Crotalaria	71	376	25	272	24	187
Desmanthus	9	313	—		—	
Desmodium	98	1531	47	2777	27	166
Dioclea	—		11	201	—	
Galactia	15	253	12	548	—	
Indigofera	80	485	17	221	25	176
Lablab	—		—		1	184
Lathyrus	—		—		7	164
Leucaena	12	683	13	193	17	174
Lotus	28	213	—		—	
Macroptilium	16	688	10	601	—	
Medicago	—		—		19	244
Neonotonia	1	291	—		1	259
Phaseolus	—		—		6	283
Pueraria	—		4	237	—	
Rhynchosia	40	367	14	453	12	140
Sesbania	38	235	—		18	305
Stylosanthes	51	2277	25	3564	14	1127
Tephrosia	75	276	20	181	—	
Teramnus	10	310	4	373	—	
Trifolium	126	1165	—		48	1598
Vicia	—		—		19	253
Vigna	56	1401	33	728	21	417
Zornia	—		17	1025	9	256
Total	(20)[b] 875	13197	(20) 349	15971	(20) 357	6769
Other genera	(161) 540	3214	(82) 209	1981	(114) 374	1571
Grand total	(181) 1415	16411	(102) 558	17952	(134) 731	8340
% 'Total' of 'Grand total'	62	80	62	89	49	81

Notes:
[a] included in 'Other genera';
[b] number of genera.
Source: Schultze-Kraft *et al.* (1993).

Documentation and information exchange.

The utility of much of the material in the collections is limited by the lack of documentation concerning its origin and its identity. Of the accessions in the Mediterranean forages surveyed by Mayer (1987) only 40% had site collection data, 20% had grid reference locations and soils information was available for less than 2%. Information concerning the site is often the most significant factor in deciding whether to select the accession for evaluation and in identifying gaps in collecting. The use of a Global Positioning System, corrected altimeter, and portable instrumentation to enter into the database site description with respect to associated vegetation, soil pH, slope and aspect will increase the efficiency of future collecting (Sackville Hamilton *et al.*, 1993).

Taxonomic uncertainties are endemic to plant collecting, and studies in species relationships, cytogenetics, floral and seed-setting biology and intraspecific diversity are needed to underpin the use of collections (Schultze-Kraft *et al.*, 1993). The standardization of descriptors, as for *Medicago* (IBPGR, 1992), facilitates the interchange of information and access to desired data.

Classification of collections

A basic problem is that morphological descriptors are adopted for their objectivity, their contribution to plant identity and their display of diversity but may have little relevance to agronomic performance. This poses the research challenge of identifying easily measured attributes which do contribute to ecological success (Pengelly & Williams, 1993). For short-lived plants flowering characteristics, seed production, seed dormancy and seedling vigour are obviously important, but mechanisms associated with the pathways of persistence are less well understood.

Some scientists have attempted to combine morphological and agronomic (M–A) attributes in the descriptors. A seminal study was the numerical analysis of *Stylosanthes* introductions (Burt *et al.*, 1971); the resulting groupings were used to rationalize the testing of representatives of this gene pool to establish their agronomic potential in different environments (Burt *et al.*, 1974; Edye *et al.*, 1975). Since the M–A groups were developed without reference to climate of origin, it is possible to test and establish the significance of differences in the climatic attributes of M–A group origins (Burt, Reid & Williams, 1976); this led to the identification of some dry tropical and subtropical zones of Central and South America which were poorly represented and required further plant exploration activity. A further example of the combined approach of morphological and agronomic descriptors is the study of *Aeschynomene* by Bishop, Pengelly & Ludke (1988).

A primary goal is to establish a core collection which is representative of the diversity within a wide collection. This meets both goals of germplasm conservation and opportunity for selective evaluation; it is often not feasible to evaluate the performance of all individuals comprising the total collection in a meaningful way. For instance, Balfourier, Charmet & Grand-Ravel (1993) used similarities between populations, geographic location and ecological characteristics of the original sites to select 114 populations of *L. perenne* which represented the variability available in 550 French wild populations. The technique of sampling populations attracts debate as to the desirable balance between the number of individuals collected per site, the number of sites, and the different characteristics of the sites represented. Marshall & Brown (1983) state '. . . the prime objective of plant exploration is to collect at least one copy of each of the different variants or alleles in a target species'. Reid & Strickland (1983) focus on the number of sites rather than the number of individuals at a site.

2.3.3 Imperatives for plant introduction

The geographical distribution of useful plants

Accessions are selected for breeding and evaluation on the basis of (i) previous experience with a species in a relevant environment; (ii) the plant taxon, especially with respect to affinities between taxa; and (iii) plant geography, especially as related to the rainfall, temperature, altitude, latitude, and soil type of collection sites (Pengelly & Williams, 1993).

Perhaps the most interesting theme in plant introduction, and one which has been largely overlooked in recent decades, is the linkage between the evolution of grass speciation and the regional climate in which evolution has occurred. William Hartley (1950) found when studying the flora of the world's regions that the lists of grass species could be arranged to show the balance of grass tribes in the list and that this balance reflected aspects of the climate of the region. Regions with a similar balance of grass tribes in different parts of the world had similar climates, and similar planted pasture species in agriculturally developed regions. A floristic list denoting presence or absence of a species was a simple tool applicable in many regions where more sophisticated data might be less available. Hartley (1954) identified five principal tribes whose balance might be used to construct an agrostological index to guide plant introduction.

The Agrosteae and the Festuceae are abundant in areas with cold winters. Their evolutionary development and species abundance (the number of species of these tribes relative to the total number of grass species present) is positively associated with latitude and altitude and negatively associated with the temperature of the coldest month. The Eragrostoideae are favoured by high temperatures but their incidence is negatively related to rainfall (Hartley & Slater, 1960). The distribution of the Andropogoneae and the Paniceae are correlated positively with mean temperature of mid-winter month and with summer rainfall for the former and annual rainfall for the latter; the Andropogoneae are more strongly developed in the eastern hemisphere whilst the Paniceae reach maximum development in the western hemisphere. The species balance of these tribes provides a biological integration of the region in which a flora is catalogued.

Hartley (1963) gives an example of the application of this approach (albeit with hindsight) to plant introduction in arid central Australia. The grass species about Alice Springs are predominantly from the Eragrostoideae (45%), Paniceae (23%) and Andropogoneae (15%). When the floras of the arid regions of the world are examined, distributions almost identical with those of central Australia are found for Yemen in Asia and for western Somalia and eastern Ethiopia in Africa. Closely similar indices are also found for Sind in Pakistan, Kathiawar in India, the Tibeti region, Sahara, northern and central Sudan, north-eastern Kenya and Namibia. The natural distribution in Africa and Asia of *Cenchrus ciliaris* (Figure 2.5) parallels the locations mentioned. *C. ciliaris* is the most successful plant introduction at Alice Springs, and its naturalization is ameliorating the trampling damage caused by tourists visiting the great rock Uluru. Williams & Burt (1982) have confirmed the broad predictive value of Hartley's concept when applied to the pastoral districts of Queensland.

Climatic homology is a guide to plant introduction success, although some scientists expect a greater success from accessions originating in a harsher environment than the one to which they are introduced; Bunting (1983) expressed the sceptical view that genotypes may be successful in the wild not so much because of their adaptation to that environment but because their neighbours are even less successful.

A good deal of research has been directed to establishing the geographical centres of species of interest (for instance, Good, 1964; Arora & Chandel, 1972; Mehra & Magoon, 1974; Harlan, 1983 a, b; Williams, 1983) and targeting regions of greatest diversity where the richest returns from

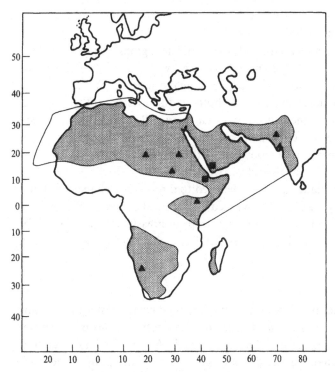

Figure 2.5. The natural distribution of *Cenchrus ciliaris* (shaded) together with regions almost identical with (■) or closely similar to (▲) Alice Springs, central Australia, in tribal composition of grass flora. (From Hartley, 1963.)

collecting occur. Schultze-Kraft *et al.* (1993) compare the paucity of existing collections with the 3800 species of tropical legumes claimed to warrant evaluation, and suggest that the majority of some 10 000 tropical grass species should have a forage potential. They list examples of under-collected areas as (1) Burma, Laos and Cambodia in south east Asia; (2) the 500–800 mm rainfall Sahelian zone and southern Africa; (3) some Central American countries; and (4) the semi-arid zone of tropical and subtropical South America (for example, the Brazilian north east and the Chaco in Argentina and Paraguay). They regard the native legume flora of tropical Africa as undervalued, whilst temperate regions with a further need for exploration are the Balkans, central and eastern Europe, including the Caucasus, western Asia, north-western China, and temperate South America, including Patagonia. Reid *et al.* (1989) refer to south east Turkey as rich in endemics and experiencing changed land use.

The exploration of habitats

Much collecting activity, especially in temperate regions, has been directed to high fertility or well managed pastures with superior productivity. Sackville Hamilton *et al.* (1993) are directing the activities of their group at Aberystwyth to different habitats, both in the interests of germplasm conservation and in the expectation that these will provide a higher incidence of novel genes; collections are therefore made from field margins, hedgerows, scrubland and moorland, paths and roadside verges, quarries, parkland, cemeteries, lawns and carparks.

Reference was made earlier to the need to direct research in accordance with the new requirements of changed agricultural systems, such as those associated with decreased inputs in European grasslands, or the need for reclamation of disturbed systems, especially in relation to the control of soil erosion.

Type of organism

It was recognized in the 1950s that the rhizobial bacteria of legume plants needed to be collected and introduced together with their hosts; the strain-specific *Lotononis bainesii* performed poorly in Australia until nodule bacteria were imported from its site of origin in South Africa. Integrated pest management may require the introduction of natural enemies not present in the new environment where a plant accession is grown. An interesting more recent concept is the advantage of introducing rumen organisms which have coevolved with the plant species in order to facilitate digestion or to alter end products to obviate toxicity. A seminal study (Jones & Megarrity, 1986) found that ruminants in Australia and Papua New Guinea lacked bacteria which avoided the degradation of mimosine in *L. leucocephala* to 3-hydroxy-4 (1H)-pyridone (DHP), which is a potent inhibitor of thyroid peroxidase, needed in the synthesis of thyroid hormones. R.J. Jones' introduction of these bacteria from goats in Hawaii radically improved the performance of cattle grazing *L. leucocephala* in Australia (Quirk *et al.*, 1988).

Plant quarantine and the introduction of weeds

Plant quarantine regulations are directed to preventing the introduction of disease, pests and weeds. Accidental introduction of weeds, especially as garden escapes from horticultural activity, have bedevilled grassland and crop productivity. Many weeds represent escapes from scientific programmes. *Pennisetum polystachion* was introduced by CSIRO to the

Northern Territory of Australia, where it is a declared weed. In the Pacific Islands this species is known as 'mission grass', since it was introduced by missionaries, whilst in Thailand the common name 'communist grass' reflects its popular attribution to the activity of Burmese communists. The *Queensland Agricultural Journal* of 1895 (p. 70) lists red natal grass (*Rhynchelytrum repens*) as a plant available for distribution to farmers; it is reported as growing well wherever it has been planted. The same journal in 1913 (p. 270) in its column 'Answers to correspondents' states: 'red natal grass is a useless weed'; subsequently, its 1895 reputation has been partially restored (Bowen & Rickert, 1979).

The introduction of woody plants with spines would now be prohibited in most countries. *Acacia nilotica* was actively planted as a shade tree along bore drains in western Queensland and has now transformed great areas of *Astrebla* grassland to woodland; some grazing properties have even been abandoned (Carter, 1994).

The threats which have arisen from the deliberate introduction of plants which have become weeds is now leading to restrictions of entry of potentially useful forage plants to plant improvement programmes. This partly arises from the difficulty of defining weed status. The naturalization of valuable herbage plants in adjacent crop lands may add to the costs of crop production. A more fundamental dichotomy of view point exists between pasture agronomists and the adherents of the primitive who seek to protect indigenous grassland communities from the invasion of foreign herbage plants, at whatever cost to the potentially enhanced stability of the ecosystem. This thrust has more force when the isolation of designated relict areas is sought. Lonsdale (1994), in seeking to identify the potential weediness of plant accessions, found his best correlation with plants described as 'useful, high performing plants'. His list of weeds introduced to northern Australia includes the legumes *Calopogonium mucunoides, Macroptilium atropurpureum, M. lathyroides, Stylosanthes hamata, S. humilis* and *S. scabra,* and the grasses *Andropogon gayanus, Brachiaria decumbens, B. mutica, Cenchrus ciliaris, Chloris gayana, Panicum maximum* and *Urochloa mosambicensis*. These plants constitute the main basis of grassland improvement in this region, have resulted in immense gains to grazing industries, have enhanced atmospheric carbon sequestration and have reduced soil erosion, and *C. mucunoides* is the only plant in the list of doubtful acceptability to farmers. Differences of viewpoint concerning the desirability of global access to plant germplasm need to be resolved if plant improvement programmes are not to be grossly constrained. Conversely, strict quarantine procedures need to be in place to

restrict the international movement of potentially invasive plants which
are unacceptable to the grazing animal.

Evaluation of introductions

Tighter public funding in many countries leads to much of the germplasm
in collections remaining unevaluated, but scientists such as Schultze-Kraft
et al. (1993) regard plant collecting as having priority for resources ahead
of plant evaluation.

Evaluation proceeds on the basis of the criteria of merit discussed at the
beginning of the chapter, and in view of the strictures suggested in the last
section concerning weediness, it is imperative that potential toxicity and the
acceptability of forage to livestock are monitored early in the evaluation
process. J.R. Harlan (1983a) considered that some of the best material collected has been found in the protection of a thornbush thicket, since selective grazing of acceptable material precluded seed production in the open;
plant collectors should accept a punctured skin in pursuit of such material.

Much of the early *ad hoc* testing by individuals or institutions has given
place to networking arrangements whereby standardized conduct of trials
is imposed on common lists of entries. This enables multilocational analysis of the data from a range of geographical sites to describe adaptability in
terms of both overall performance and the responsiveness of accessions to
environmental variation, as in the RIEPT programme of CIAT in central
and south America (Keller-Grein *et al.*, 1993). Elite accessions can then
pass through the series of more critical evaluations leading to commercial
seed multiplication and release (Ferguson, Vera & Toledo, 1989).

2.4 Approaches to plant breeding

2.4.1 The decision to breed

The preferential allocation of resources to plant breeding rather than to
plant introduction, selection and evaluation may be misguided unless it
follows a long period of collection and research or represents a sole option
in responding to a severe agronomic problem. D.F. Cameron (1983) suggested: 'experience . . . shows clearly that plant introduction should be the
primary source of yield improvement in pasture species'. J.R. Harlan
(1983a) reported: 'I spent as much as 10 years making modest progress only
to acquire new accessions several times as good as the best of my produc-
tions. I worked with some species for years before I found that other species
were more appropriate'. In Queensland there have been serious institu-

tional commitments to forage plant breeding for *c.* 45 years; of the principal forage species planted three legumes were bred and 20 grasses and legumes resulted from simple plant introduction and selection.

It is also desirable to know when the main gains have been won and the particular programme might be closed. E.M. Hutton had a notable success through the hybridization of two Mexican introductions of *M. atropurpureum* which produced transgressive segregates with a marked intensification of the stoloniferous character; at the F4 stage the three best selections were combined and released as the cultivar Siratro in 1960 (Hutton & Beall, 1977). After a further 20 years of intensive breeding directed to improvement of several characters eight elite lines were chosen for widespread regional testing; none of these proved superior to the original cv. Siratro. Subsequently, the susceptibility of Siratro to rust (Jones, 1982) reawakened further breeding activity.

The opportunity for gain from breeding is both appropriate and necessary for varietal stability in outbreeding temperate plants such as *Lolium, T. repens, T. pratense* and *M. sativa*; special breeding techniques are required for self-fertilized plants and for the apomictic grasses so prevalent in the tropics. The longer history of research in temperate regions is reflected in a greater emphasis on breeding rather than introduction, but examples have already been given in this chapter of recent innovations based on introductions of *T. repens* and *Lolium* breeding has benefited similarly (Jones & Humphreys, 1993). The annual advance in DM yield of *L. perenne* due to breeding in Belgium, Netherlands and UK over a 25-year period to 1990 was estimated as 0.5% (Van Wijk, Boonman & Rumball, 1993); the preceding decades were less fruitful. A larger claim was made for the diploid *L. perenne* cv. Aberelan (Institute for Grassland and Environmental Research, 1994), which was considered to represent an annual gain over the previous 25 years of 0.88% for annual DM yield and 1.85% for early spring growth. Advance in yield performance of *T. repens* was estimated as 0.4% per year since the 1930s (Caradus, 1993).

Objectives in plant breeding in some programmes have covered all of the criteria of merit discussed, but many scientists focus on disease and insect resistance as themes with a special resonance for breeders.

2.4.2 Techniques of plant breeding

At the beginning of this chapter reference was made to the knowledge for genetic advance which was in place in 1952. This brief account enumerates the main techniques of relatively recent and current practice.

Selection and evaluation.

Selection requires variability, which is available in open-pollinated species and in wide collections of self-fertile species. In the 1950s and 1960s it was fashionable to increase variation by radiation-induced mutation (Brock, 1971), which occurs randomly, but as few gains were won this approach mainly passed into desuetude.

The discovery that repeated clonal regeneration led to somatic variation in *Lolium* (Hayward & Breese, 1968) awakened fears in some scientists that the ghost of the Russian geneticist Lysenko, who affirmed that environmental influences caused heritable somatic change, was stalking the corridors of the Welsh Plant Breeding Station. However, the phenomenon of somaclonal variation was confirmed by many scientists (for example, Godwin, Cameron & Gordon, 1990); it has not led yet to any significant cultivar release.

The major options (Bray & Hutton, 1976) are mass selection, pure line selection, and hybridization followed by pedigree selection in which data on parental origins are maintained. Michaud, Viands & Christie (1993) found in comparing selection procedures of *M. sativa* based on polycross progeny performance with those based on topcross progeny performance that the latter gave a higher heritability estimate and a greater genetic advance for yield. Desirable characters in herbage plants, such as yield, persistence and nutritive value are usually under polygenic control, which restricts the rate of improvement breeders can achieve and the effectiveness of backcrossing. In seeking successful adaptation to environment scientists recognize two mechanisms: individual buffering in homogeneous populations, in which individual genotypes absorb or avoid a range of stresses, or population buffering, in which the balance of cohabiting genotypes which may exploit different ecological niches reflects an optimum frequency in response to selection pressures (Clements *et al.*, 1983). The selection procedures adopted in these circumstances often reflect the mode of reproduction.

It is desirable that field selection occurs across the range of target environments, and most successful herbage varieties have a wide range of adaptation, unless the breeding programme was directed to filling a gap for an unusual, narrow ecological situation. Testing usually commences in a spaced plant configuration, which is appropriate for monitoring some characters such as disease and pest resistance and survival of freezing. The applicability of data from spaced plants to sward conditions for other characters is debated; for example improved nutritive value of *Dactylis glomerata* under spaced plant conditions was (Buxton & Lentz, 1993) or was not (Cooper & Breese, 1980) found to be applicable in swards. Synchronous growth after cutting *M. sativa* was favoured by selection in dense swards (Rotili, 1993). The early application of grazing to breeding material leads to gains (Bouton *et al.*, 1993b for *M. sativa*), but the defoliation system needs to be appropriate for farm practice, as illustrated by the development of large-leaved cultivars of *T. repens* adapted to cutting for conservation and of small-leaved cultivars successful under close grazing by sheep (Evans, Williams & Evans, 1992).

Hybridization

J.R. Harlan (1983b) suggested that: 'the art and science of plant breeding consists, to a large extent, in finding parents whose offspring perform better than they do'; the best combinations need to reflect optimum levels of divergence. Serendipity may appear in the successful choice of parents. The combination of overwintering ability and early spring growth was unexpectedly evident in accessions of *Cynodon dactylon* from South Africa; Harlan (1983a) speculated that these traits may have evolved during the Pleistocene era when South Africa was much colder than now.

Cytogenetic studies to describe chromosome configuration, to enumerate level of ploidy and to cast light on the mode of reproduction provide the basis for some hybridization programmes, especially where wide crosses are involved. The description of the mode of reproduction indicates the degree of self-fertilization or of self-incompatibility in outcrossing systems, the amount of inbreeding depression, and the presence of sexuality or apomixis (Burton, 1983). This is allied with studies of floral biology which suggest how synchronous crossing systems may be developed. The discovery and transfer of cytoplasmic male sterility used in conjunction with elite male parents provides high-yielding hybrids (Suginobu *et al.*, 1993). Somewhat similarly the combination of a sexual female parent with pollen from qualitative apomictic plants fixes heterosis and the genome in the F1 generation (Bashaw, Voigt & Burson, 1983). Self-sterile plants such as *M. sativa* are more readily bred than self-compatible plants whose flowers require emasculation. In outbreeding plants such as *Lolium* synthetic varieties are constituted with a sufficient number of nucleus plants to avoid inbreeding depression, and seed production of varieties may be limited to four or five generations to avoid genetic drift (Breese, 1983).

Induced polyploidy and wide crosses

An early example of ingenuity in overcoming ploidy and apomictic barriers to crosses was provided by Burton & Forbes (1960); chromosomes of sexual diploid plants of *Paspalum notatum* were doubled and these provided the induced tetraploid female parents which were crossed with apomictic autotetraploid *P. notatum* cv. Pensacola. This released new variation; apomixis was found to be controlled by a very few recessive genes. Interspecific crosses between the sexual *Brachiaria ruziziensis* and the obligate apomictic *B. decumbens* and *B. brizantha* were also achieved by inducing autotetraploid *B. ruziziensis* as the female parent in a programme directed to increasing resistance to spittlebug (Homoptera: Cercopidae) and adaptation to low soil fertility; apomixis appeared to be under monogenic control (Do Valle, Glienke & Leguizamon, 1993). Quantitative apomixis poses greater problems for breeders; an alternative approach to the above is to attempt fertilization of unreduced eggs (2n + n hybridization), which frequently occurs within agamic complexes such as *Cenchrus–Pennisetum* (Hussey *et al.*, 1993).

Two perennial themes in European, USA, Japanese and New Zealand grass breeding have been the hybridization of *Lolium perenne* × *L. multiflorum*, and attempts at intergeneric *Lolium* × *Festuca* crosses. The former seeks the complementation of specific characters, such as the rapid establishment and high yield of conservation cuts of *L. multiflorum* with the high tiller density, persistence, resistance to treading and winter hardiness of *L. perenne* (Breese, 1983). As these programmes have progressed it has appeared that the use of tetraploids, with tetrasomic inheritance, as well as preferential pairing between homologous chromosomes, reduces segregation so that the production of four or five generations of seed multiplication does not give a drastic reduction in hybridity (Jones & Humphreys, 1993).

Humphreys, Humphreys & Thomas (1993) consider that in breeding for improved resistance to environmental stress there are advantages in introgressing a limited number of genes from *F. arundinacea* to *L. multiflorum* rather than in combining complete genomes. One technique for introducing the drought resistance of *F. arundinacea* is to produce pentaploid hybrids by crossing tetraploid *L. multiflorum* (2n = 28) with hexaploid (2n = 42) *F. arundinacea* and culturing the embryos 16 days after pollination. The hybrids may then be used as male parents in backcrossing twice to diploid *L. multiflorum* in which isozyme marker alleles from *F. arundinacea* are retained, and selection pressure for drought resistance is maintained by the use of a rainout shelter. Intergeneric hybrids of male-sterile *L. multiflorum* and *F. arundinacea* have been shown in Japan to exhibit superior voluntary intake and lower levels of acid-detergent fibre relative to *F. arundinacea* (Suginobu *et al.*, 1993) but the problem of inadequate seed production from these hybrids remains.

The failure of endosperm production after interspecific pollination in temperate legumes was overcome by Williams & Williams (1983) by sterile culture of a hybrid embryo after transplantation into a nurse endosperm taken from a normal ovule. These and other sophisticated techniques are enlarging the horizon of possibilities for the success of wide crossing.

2.4.3 Molecular biology

The first paper on molecular biology at an International Grassland Congress was delivered in 1985 by I.K. Vasil of the University of Florida. Whilst Vasil (1985) referred to the current possibility of attempting transfer of specific genes from other plants or organisms to herbage plants by genetic transformation, the main thrust of the paper dealt with the regeneration of plants from cell suspension cultures, callus cultures and protoplasts. At this stage tissue culture was well established, especially with horticultural plants, and provided rapid multiplication and maintenance of disease-free material having genetic integrity.

The rationale for genetic engineering has three central assumptions. The first is that the structure of plant variation can be understood in the new terms of its basis in the molecular biochemistry of gene action. The second is the creation of novel characters unavailable in traditional plant breeding practice. The third (Peacock, 1993) is that: 'the genetic engineer introduces a short segment of the coded tape, essentially a single gene construct, into the pre-existing genome of some agricultural species, whereas a breeder has the much more difficult task of bringing in the required segment via a whole chromosome set through hybridization and then having to work hard to recombine the required gene away from all of the non-required genes introduced in the cross'. Peacock continued optimistically that this shortened the breeding timetable.

This brief introduction to the topic outlines three types of primary activity: (1) the identification and isolation of gene codes; (2) the transformation of the target plants by incorporation of foreign gene codes; and (3) the regeneration of the transformed plants and the analysis and optimization of transgenic expression. Examples are then given of transformations directed to the improvement of herbage nutritive value and resistance to disease and pests.

The identification and isolation of gene codes
Genetic linkage identified by classical breeding methods and isozyme markers are entrenched approaches. Restriction fragment length poly-

morphisms (RFLP), in which an endonuclease enzyme is used to cut genomic DNA into fragments at sites recognized by the enzyme, represented an advance; the association of the fragment with an organism trait facilitated identification. This technique was applied to fungi as well as plants; for example, the existence of two distinct populations of *Colletotrichum gloeosporioides* was confirmed following DNA extraction and restriction enzyme analysis by polymorphisms in ribosomal DNA repeats and after hybridization with low and high copy random genomic probes (Braithwaite, Irwin & Manners, 1990). A more recent class of molecular marker based on polymerase chain reaction (PCR) amplification of genomic DNA using random sequence primers (RAPD, Timmerman & McCallum, 1993) synthesizes many copies of the target sequence which can then act as template for further rounds of amplification; an accumulation of the product is then analysed by electrophoresis or used in other recombinant DNA methods.

A specific example of gene encoding and transformation is the research directed to the enrichment of *Trifolium subterraneum* with high S-containing amino acids (Khan *et al.*, 1996). It is known that growth, reproduction and wool production in sheep is limited by the amount of methionine and cysteine reaching the abomasum, since they respond to these as supplements. Sunflower seed albumin (SSA) is rich in methionine and cysteine (Tabe *et al.*, 1993) and a chimeric SSA gene was constructed by the addition of a 5' region that ensured leaf expression, and a sequence ensuring that retention of protein in the endoplasmic reticulum was favoured. The protein-coding region of this SSA gene was further modified by the inclusion of a nucleotide sequence encoding additional amino-acids (TSEKDEL) upstream of the stop codon. A gene promoter from the cauliflower mosaic virus 35S RNA gene was used and the termination region was from the same gene. The SSA gene was inserted into the binary vector pTAB IO, which contained *Agrobacterium tumefaciens* borders flanking a chimeric gene encoding resistance to phosphinothricin (PPT), and was transferred to *A. tumefaciens* AGL I; the integrity of the gene construct in *A. tumefaciens* was confirmed by restriction enzyme digestion. This system is illustrated in Figure 2.6.

Transformation of target plants by incorporation of foreign gene codes

In this example *A. tumefaciens* harbouring the recombinant plasmid was grown in Luria broth, and was used for inoculation of hypocotyl explants from *T. subterraneum*. An alternative to the use of explants is the transformation and regeneration of protoplasts (Hodges, Rathore & Peng, 1993, to whom this section is indebted). Polyethylene glycol may be used for facil-

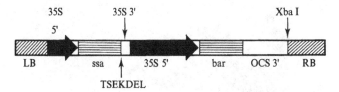

Figure 2.6. Schematic structure of the two chimeric genes transferred into *T. subterraneum*. DNA encoding the sequence of amino acids (denoted TSEKDEL) was added to the 3' end of the SSA protein-coding region to ensure protein retention in the endoplasmic reticulum. The promoters (denoted 35S 5') of the two genes appear as black arrows, and the terminators (35S 3') appear as clear boxes. The left and right borders (LB and RB) indicate the boundaries of the T-DNA. The XbaI restriction site was used to digest genomic DNA for Southern blotting. (From Khan *et al.*, 1996.)

itating the uptake of DNA, but its efficiency for grasses is low. Another approach is electroporation-mediated transformation which uses the concept of electric field-mediated membrane permeabilization to assist the movement of DNA molecules into plant protoplasts, but there are often problems of subsequent regeneration.

Microprojectile bombardment devices have been developed for transforming cells within tissue masses; these use a gunpowder explosion, compressed gases or an electrical discharge. Tungsten or gold particles about 1 mm in diameter are employed, and embryonic callus tissue or suspension cells are used as targets. Transformations mediated by micro-injection have been mainly used in animal studies but have been successful in *M. sativa* (Reich, Iyer & Miki, 1986). The development of novel transformation techniques is evolving rapidly.

Regeneration and the effectiveness of transformation

In the example with *T. subterraneum* (Khan *et al.*, 1996) the inoculated explants were transferred to a cocultivation medium for 7 d, the root segment was cut and discarded, and the secondary explants were placed in a regeneration medium, excised, and rooted. RNA was isolated from the primary transgenic plants and from two successive generations the level of SSA mRNA was monitored using Northern blot analyses, and the T_2 generation plants were all found to be resistant to PPT; these findings indicated successful transformation. Immunogold labelling in the leaves of transgenic *T. subterraneum* showed that SSA was localized in large electron-dense inclusions within a lumen delimited by a ribosome-studded membrane (Figure 2.7). The albumin formed up to 0.3% of total extractable

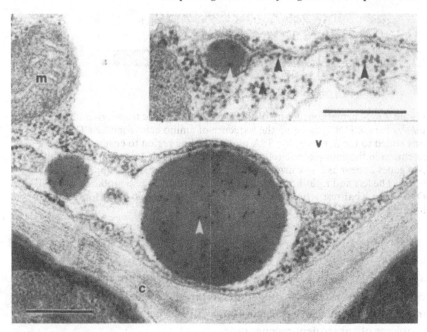

Figure 2.7. Electron micrograph showing immunogold localization of SSA within a xylem parenchyma cell in the leaf of transgenic *T. subterraneum*. The insert shows a mesophyll cell with an SSA-containing, electron-dense deposit within a rough endoplasmic reticulum cisterna. (v = vacuole; m = mitochondrion; c = cell wall; white arrows = gold particle in an electron-dense body; black arrows = cytoplasmic and membrane-bound ribosomes. Scale bar represents 0.5 μm.) (From Khan *et al.*, 1996.)

protein, increasing to 1.3% in older leaves, indicating preferential retention with age. These levels are at the bottom end of expectations for positive responses in sheep and further crossing is being undertaken to intensify the expression of SSA in *T. subterraneum* leaves.

Much laboratory technology is directed to devising plant regeneration procedures specific to particular target plants, and elaborate protocols are often necessary to ensure callus initiation and regeneration. These are also directed to obtaining somaclones of intergeneric hybrids with restored plant fertility (Hesky *et al.*, 1993) and to generating variability in sterile triploids such as *Digitaria decumbens* (Cheng, 1993) from embryogenic callus formation and regeneration from immature inflorescences.

Increased nutritive value

Since high lignin content reduces nutritive value, as discussed early in this chapter, a significant thrust of plant improvement is directed to antisense

RNA technology which would not add a new function but which would reduce the expression of the genes for lignin biosynthesis. The terminal enzyme leading to this synthesis is peroxidase, and ribozymes or 'gene shears' can be targeted together with antisense technology to reduce peroxidase level (McIntyre *et al.*, 1993).

A different approach is to modify the organisms in the rumen to increase the efficiency of digestion. Anaerobic fungi such as *Neocallimastix patriciarum* contain genes coding for relatively high-level activity of exo-ß-1,4–glucanase, which is involved in cellulose hydrolysis (Orpin & Xue, 1993), and of other enzymes involved in plant cell wall breakdown. Attempts are being made to insert fungal genes, which can already be inserted into the bacterium *Eshericha coli*, into rumen bacteria.

Resistance to disease, pests and herbicides

Programmes directed to the management of *Colletotrichum gloeosporioides* (Chakraborty *et al.*, 1996) were mentioned earlier. Other efforts are directed to anti-fungal protection through the incorporation of high levels of chitinase, or through ribosome-inactivating proteins such as Ricin. Considerable success has been achieved in the control of virus diseases with a gene expressing the coat protein of tobacco mosaic virus (TMV) (Fitchen & Beachy, 1993). It is based on the finding that once a plant is infected with a virus strain it becomes resistant to closely related strains. This approach of coat-protein mediated resistance causes plants to synthesise the protein that, along with TMV RNA, normally forms virus particles and therefore accords resistance to TMV challenge. It has been extended to many virus groups.

Research in insect resistance has used genes coding for insecticidal proteins from the bacterium *Bacillus thuringiensis*, and for inhibitors of protein digestion (White *et al.*, 1993). Inhibitors of carbohydrate digestion, lectins, feeding deterrents and insect peptide hormones constitute other possibilities.

Finally, resistance to herbicides, which has been instanced as a marker for transgenic success, either is sought to facilitate pasture and crop management or arouses concern because of enhanced potential for weediness.

Molecular biology has become such an exciting and rapidly developing field and has attracted such a wealth of community and private investment that it is necessary to reiterate an earlier statement that by 1995 it has generated no impact on grassland practice. The affirmation of Frey (1992) that the core of plant breeding will remain very much intact during the next decade might be noted; the most expensive part of cultivar development is the field and laboratory testing for quantitatively inherited traits and the most extensive and long-term tasks will be the continuing field testing of candidate varieties for their zones of adaptation and consumer acceptability. Genetic engineering at the molecular level might therefore be regarded as one component of a multidisciplinary breeding programme. The lawyer might also be considered an additional player.

2.5 Plant variety protection

The culture of germplasm access and exchange and the role of public plant breeding, with its former practice of open release of new cultivars to seed growers and farmers, have changed radically since the 1937 International Grassland Congress. On the one hand concern has been expressed that multinational companies would deny the farmers of the world the opportunity to use elite germplasm if they conceived their interests to be threatened; on the other hand a sinister association of crop protection practices involving the use of hard chemicals is seen to be tied to varieties requiring intensive management. Access to indigenous plant material has also been restricted by some nations seeking to protect their ownership of germplasm originating in that geographical region. Community acceptance of transformed plants is a further imponderable.

The introduction of plant variety rights, which for example has been operative in the USA since 1930 for asexual species and since 1970 for sexual plants (Frey, 1992), has been the major factor in changing plant improvement programmes. It has generated immense investment, both private and public, in plant breeding and in basic research in molecular biology. This has arisen since ownership of varieties and of intellectual property was available, protecting and generating investment.

Although plant variety rights are often viewed negatively from the perspective of ownership, positive aspects should also be given their due. Regulation of varietal availability is also linked to genetic integrity; a wholly free market in seeds is often linked to inferior genetic performance and the transmission of weeds, pests and disease. There are benefits to developing or receiving countries in terms of the availability of better cultivars and of an improved infrastructure for the production and distribution of quality seed. Market orientation also leads to an emphasis on utility in farm practice, since unsuccessful varieties cease to be sold. The traffic of ideas and experience between farmers, seed retailers, extension agents and breeders has been generated in some countries to the great advantage of the relevance of breeder activity (Laidlaw & Reed, 1993; Van Wijk *et al.*, 1993).

2.6 Conclusion

Continuing advances in the provision of elite germplasm for diverse situations directed to the improvement of sustainable land use, whether for farming or other purposes, depend upon the balanced allocation of scarce resources to plant collection, habitat protection, evaluation of plant introductions, molecular biology and plant breeding in its widest sense. These advances also depend upon investment in crop physiology, taxonomy, seed technology, crop protection, and animal nutrition. As agricultural science becomes increasingly specialized there is a shortage of scientists with the background to evaluate agricultural problems from a necessary breadth of disciplinary and field experience; can the laboratory worker in the molecular genetics of *M. sativa* formulate criteria of merit which place this plant in a context in which crop and forage production are integrated in a farming system?

The vision of global sharing of forage germplasm persists. Reid (1993) wrote: 'We are all familiar with the importance of, for example, West African *Andropogon gayanus* in Brazil, Brazilian *Stylosanthes scabra* in Australia, Spanish *Trifolium repens* in New Zealand, Kenyan *Cenchrus ciliaris* in Mexico and Italian *Medicago sativa* in Argentina. Imagine this concept applied in the near future, when Australian *Glycine* spp. are revegetating degraded pastures in India, Mexican *Leucaena* spp. are providing browse plants in Ethiopia and African *Trifolium* spp. are supplying high quality forage through the tropical highlands of the world.'

This chapter has primacy amongst the topics which provide the bases for grassland improvement. Genetic advance is also more readily adopted than other variations in land use practice. The incorporation of elite seed or planting material in land systems is the single innovation most effective in contributing to gains in productivity and in environmental protection.

3

The nitrogen economy of grasslands

3.1 Introduction

Nitrogen availability is the primary mineral constraint to plant and animal production from grassland. The supply of other nutrients, such as phosphorus, limits plant response to N, but on a global basis N deficiency is the most significant nutrient factor which determines the level of animal output and the stability of the ecosystem. It is also a production factor which is within the power of the farm manager to modify.

Other edaphic themes occupy the grassland scientist: salinity and mineral toxicities, the balance of nutrient requirements, and the stability and structure of the soil. Nitrogen economy (Whitehead, 1995) is selected as the theme of this chapter because of the close positive relationship between growth and soil N (see for example, Graham *et al.*, 1981), and because of the controversies which have occupied grassland scientists in the past 50 years. These have centred about the reliance on the legume for biological N fixation and alternatively on N fertilizer produced from fossil fuel, and more recently about the atmospheric and stream pollution arising from N in agricultural systems.

The Fourth International Grassland Congress

At the IV International Grassland Congress, held in Aberystwyth, Wales, in 1937, the President R.G. Stapledon affirmed: 'No grassland is worthy of the name, and indeed is hardly worth bothering with, unless a legume is at work. Find or breed the right legume for every corner of the world and you have tolerably good grassland in every corner of the world. Make the conditions suitable for the legume and manage the sward to favour the legume as well as to feed the animal, and everything else will be easy – the battle will be won'. This exhortation has been persistently echoed by grassland scientists, especially in New Zealand, Australia and California.

N fertilizer

Other emphases were evident at this Congress. There had been a long tradition of studies about the return of animal excreta to grasslands, but although the application of chemical fertilizers supplying N to grasslands had almost no place in farm practice, there were several papers about N fertilizers.

In Scotland Heddle & Ogg (1937) applied 21 kg N ha^{-1} and noted that this suppressed clover growth, whilst in Sweden Osvald (1937) referred to '. . . a reaction against the strong propaganda for liberal nitrogen dressing of grassland', the depression of clover, and '. . . a deterioration of the crop'. The inhibitory effect of nitrate on nodulation was well known (Fred & Graul, 1916). Giöbel (1937) was adventurous with respect to N level, and tested applications of up to 217 kg N ha^{-1}, which promoted early spring growth, so that commencement of grazing in Sweden might be advanced by 5 to 8 d.

The results of the previous cutting experiments were reinforced by measurements of animal production from steers grazing mixed pastures in Connecticut, USA (Brown, 1937). The application of superphosphate, lime and potash increased production by a factor of 2.2, whilst the additional application of 62 kg N ha^{-1} as split dressings in April and June increased production by a factor of 3.5; application in August was less successful.

Clover

Many scientists at this time focused on the nutrition of clover, the beneficial effects of superphosphate and lime (Hanley, 1937; Heddle & Ogg, 1937) and on the need for potassium on some soils (Giöbel, 1937). There were also efforts to devise management systems which combined the use of N fertilizer with the retention of clover in the sward (Frankena, 1937).

A.I. Virtanen (1937) reported the excretion of N as amino-acids from the root nodules of pea plants grown in sterile culture in Helsinki, Finland. The phenomenon appeared to be most rapid during the early growth of the plant, and depended on the effectiveness of the rhizobial strain used for inoculation.

3.2 The development of N fertilizer practice

3.2.1 N fertilizer use
By the time of the VI International Grassland Congress the horizon for N requirement of grassland had lifted, and Mulder (1952) referred to the need

for grassland uptake of 300 kg N ha^{-1} to produce a yield of 10 t ha^{-1}, which was readily attainable in the Netherlands. William Davies, the Director of the Grassland Research Institute at Hurley, UK, suggested that to attain high production 'we must face the loss of clover and accustom ourselves to a regime of intensive pasture production wherein little or no reliance is placed on clovers' (Davies, 1960).

The expansion of industrial capacity in many countries after World War II led to the availability of cheaper combined N, and initially this was mainly applied to crops. In the Netherlands average application to permanent grassland increased from *c.* 45 kg N ha^{-1} yr^{-1} in 1950 to *c.* 150 kg N ha^{-1} yr^{-1} in 1964 when the level of application to field crops was much less than this (Van der Molen & t' Hart, 1966). In other European countries and the USA the move to intensification of grassland production through the N fertilizer route developed later; in the tropics and subtropics N fertilizer had a minor role at this time in grasslands used for dairying.

The thrust to increased use of fertilizer N on grassland gained further momentum until the energy crisis of 1973 increased prices, but there was a sustained growth into the 1980s: in Europe grassland fertilization in 1980 varied from *c.* 28 kg N ha^{-1} in Ireland to *c.* 236 kg N ha^{-1} in the Netherlands (van Dijk & Hoogervorst, 1982). This was matched by a decreasing content of clover in European swards (Picard & Desroches, 1976; Crespo, 1982).

3.2.2 The responsiveness of grassland to N fertilizer

Plant response to increased N is modified by weather conditions, soil temperature, antecedent soil moisture, the structure, location and fertility of the soil, and the pasture species and its management.

Scientists from the Netherlands devised a simple technique for displaying the growth, N uptake and apparent N recovery of swards in relation to N application. This is illustrated for a *Lolium perenne* sward established on a previously cropped polder in the Netherlands containing a clay soil with a low organic matter content (Figure 3.1, Sibma & Alberda, 1980). Growth during a 168-d growing season reached a maximum yield of *c.* 17 t ha^{-1}; the linear response phase extended to higher levels of N application in swards cut frequently at intervals of 14 d, and this treatment gave lower yields than treatments which were cut more infrequently (see quadrant II). N uptake increased with increasing DM yield (see quadrant I) and herbage N concentration usually fell between 1.6 and 4.0%, increased with rising N uptake, and was higher with more frequent cutting, as is discussed in chapter 4. N uptake also increased to the highest level of N application, 800

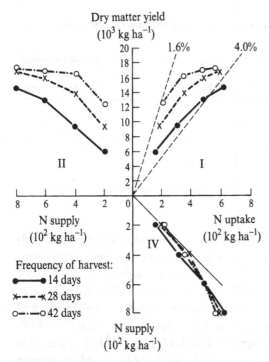

Figure 3.1. Effect of N application rates and frequency of harvest of *Lolium perenne* on the relation between N uptake and dry matter yield (quadrant I), N application rate and dry matter uptake (quadrant II), and N application rate and N uptake (quadrant IV). Thin dashed lines inquadrant I indicate N concentration. (From Sibma & Alberda, 1980.)

kg N ha^{-1} (see quadrant IV), and was little affected by cutting frequency. The thin solid line in this quadrant indicates an equality of N uptake and N application, so the apparent N recovery was high, especially at the lower levels of N supply; the contribution from native soil N was not estimated in this particular series, which lacked a zero application treatment.

The plant response to N fertilizer is greater under conditions of high radiation and is closely linked to soil moisture supply (Cowling & Clement, 1974; Henzell *et al.* 1975). Both drought and excessive moisture supply limit the responsiveness of the sward (Van der Molen & t'Hart, 1966). In the Netherlands well-drained peat soils were less responsive than sandy soils (de Boer, 1966).

Pasture conditions in the warm humid tropics provide a radical contrast to these examples. Studies in Hawaii and in Puerto Rico were seminal in drawing attention to the high levels of primary productivity, the great

Table 3.1 *Response characteristics of* Pennisetum purpureum *to variation in N supply in Puerto Rico*

N fertilizer level (kg ha^{-1})	Yield (t ha^{-1})	Response (kg DM kg N^{-1})	N%	Apparent N recovery (%)
0	34.1	–	0.9	–
224	47.8	61	0.8	35
448	63.3	65	0.9	56
896	84.7	57	1.1	71
1344	77.9	33	1.3	54
2240	85.9	23	1.5	46

Source: Vicente-Chandler *et al.* (1959).

responsiveness of growth to augmented N supply, the need for high levels of N application if growth potential were to be expressed and the problems of protein dilution and of maintaining forage quality, especially under conditions of lenient use. This is illustrated by a study at Rio Pedras, Puerto Rico on an alluvial clay loam planted to *Pennisetum purpureum* and well fertilized with P, K and lime (Vicente-Chandler, Silva & Figarella, 1959). The annual production of forage cut at 90 d intervals (Table 3.1) increased from 34.1 t ha^{-1} at zero N to 85.9 t ha^{-1} at 2240 kg N ha^{-1}; DM production was near-maximal at the 896 kg N level. The responses of 61–65 kg DM kg^{-1} N applied in the zero–448 N range might be compared with the values of 10–25 kg DM kg^{-1} N expected from grasses growing in temperate regions.

Table 3.1 indicates good apparent recovery of N up to the 896 N level and continuing uptake of N to the highest level tested. On the other hand, although N concentration in plant tops increased with increasing N application level, the nutritive value of the forage produced was unsatisfactory; the 90 d cutting interval maximized the interception of solar radiation by this grass with a C$_4$ photosynthetic pathway, but increased the lignin content and decreased the concentration of P, Ca, Mg and K. In the same study, cutting *P. purpureum* at 40 d intervals produced average N concentrations in the shoots of 1.3, 1.6 and 2.8% in the N0, N448 and N2240 treatments respectively, and highest DM production was limited to a mere 35.4 t ha^{-1}.

3.2.3 *Seasonality of production*

The role of N fertilizer in extending the grazing season and in maintaining continuity of forage supply was recognized by scientists at the 1937 IV

International Grassland Congress. It was known that in Europe and North America cold soil temperature restricted N mineralization, and the application of a readily available N source stimulated grass growth in early spring (Blackman, 1936). Many studies then followed which quantified this response in terms of environmental variables, N source and level, and plant species and which drew attention to subsequent compensatory effects later in the season.

Thus, at Hurley, UK, Cowling (1966) found that applying nitro-chalk to *L. perenne* – *T. repens* swards in mid-February advanced the date when grazing might commence by 2–14 d, depending on weather conditions, and gave a yield response of 15–17 kg DM kg^{-1} N. He also noted depressions of yield at the time of the second cut if inflorescence development were impaired and at the third cut when summer growth of clover was reduced. In subtropical Australia grass growth in the autumn was greatly limited by N availability at the end of the main growing season (Henzell & Oxenham, 1964).

A philosophy therefore developed (for example Vélez & Escobar, 1970; Murtagh *et al.*, 1980) that herbage supply and animal requirement for forage might be brought more nearly to synchrony by the judicious application of N fertilizer. In the tropics and subtropics the recognition that forage has a scarcity value and is well utilized in periods of shortfall led managers to accept a lesser N response of *c.* 10–25 kg DM kg^{-1} N relative to a greater N response of perhaps 30–50 kg DM kg^{-1} N which might be attained in the main growing season; the latter might contribute to a forage surplus, unless conserved.

The environmental constraints to the promotion of out-of-season growth are illustrated by a study in coastal northern New South Wales of the spring and early summer growth of *P. clandestinum* or *Setaria sphacelata* var. *sericea* swards (Murtagh, 1975). The responses to an application of 112 kg N ha^{-1} made between August and December were analysed in terms of climatic and soil variables. Mean night temperature increased from *c.* 10 °C to 22 °C over the period and was the temperature variable most closely allied to both DM response and apparent N recovery. At 10 °C night temperature DM response four weeks after N application was 40% and N recovery was 60% respectively of the values recorded at 20 °C night temperature. Effective rainfall, defined as rainfall minus runoff, and the average soil water potential (0–15 cm depth) over the period provided the best moisture indicators of DM response (Figure 3.2) and of N recovery. Both moisture indicators were needed to describe pasture response, and an abrupt transition from dry to wet conditions soon after topdressing enhanced the

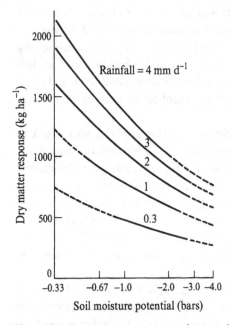

Figure 3.2. Dry matter response at four weeks after N application at varying levels of soil moisture potential and effective rainfall, (assuming 16 °C mean night temperature) in northern New South Wales. Solid lines indicate the actual experimental range of values. (From Murtagh, 1975.)

response to N fertilizer relative to the response during continuing wet conditions. Pasture yield four weeks after fertilizing increased by 300 to more than 2000 kg DM ha⁻¹ at intermediate night temperature, according to moisture conditions (Figure 3.2). The utility of N application in this environment therefore hinges on the pattern of spring rainfall (Murtagh, 1977), since monthly temperatures are relatively predictable.

3.2.4 Response of plant species

Plants vary greatly in their adaptation to level of N supply. This is evident when the performance of monospecific swards is compared, and stronger contrasts appear when plants are grown in mixed swards. Frankena (1937) noted an increase in preferred species such as *L. perenne* when swards of diverse botanical composition received N fertilizer.

The competitive relations of grasses on a tropical tableland pasture at Millaa Millaa, Queensland (lat. 18° S, 2620 mm annual rainfall) as modified by N supply are illustrated in Figure 3.3 (Gartner, 1969). The soil was

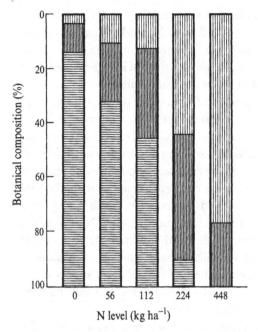

Figure 3.3. Botanical composition of a tropical tableland grass sward after 3 years at different levels of by N application (▤ *Axonopus affinis*, ▦ *Paspalum dilatatum*, ▥ *Pennisetum clandestinum*). (From Gartner, 1969.)

a loamy sand and N fertilizer was applied as urea in split applications to cut swards of mixed composition. *Axonopus affinis* is well adapted to low fertility conditions and dominates many run-down pastures in the tropics and subtropics. In this study it disappeared after three years of treatment at the N448 level and was a minor component of the sward at N224, as depicted in Figure 3.3. *Paspalum dilatatum*, which has been widely sown as a dairy pasture grass, occupied an intermediate position whilst *Pennisetum clandestinum* dominated the sward under high N supply. These effects were even more pronounced under grazing (Mears & Humphreys, 1974a). There are other reports of *P. clandestinum* maintaining a higher N concentration in the shoots than other tropical grasses, and showing a greater response to N than might have been expected from its growth capacity (Wilson & Haydock, 1971).

Efficient response to nutrient availability arises from superior capture of the nutrient by the grass root system and its delivery to the shoots, and/or greater shoot production per unit of nutrient uptake, as mentioned in chapter 2. The dietary value, as expressed positively in a higher protein content or negatively in terms of high NO_3 level, are further factors of interest. Cowling & Lockyer (1967) found at Hurley, UK that *Dactylis glomerata* cv. S37 and *Phleum pratense* cv. S48 showed N uptake which was superior to five other temperate grasses. On the other hand, the efficiency of shoot DM produced per unit of N uptake was greatest in *L. perenne*, which averaged 41 units, relative to 34 units for *Agrostis tenuis*.

Early in this chapter the N response of *L. perenne* in the Netherlands provided an extreme contrast with the performance of *P. purpureum* in Puerto Rico. The generic differences between tropical C_4 and temperate C_3 grasses were better compared in a glasshouse study in Brisbane (Wilson & Haydock, 1971) in which nine C_4 grasses and ten C_3 grasses were grown in varying conditions of N and P nutrition but were harvested at similar DM yields. This study confirmed many field observations which suggest that tropical grasses grow better than temperate grasses at low N supply, and accumulate lower N concentration in shoots than temperate grasses. The lower protein contents reported for tropical grasses appear to be associated with more rapid growth, greater retention of assimilate in shoots rather than roots, and a higher proportion of structural and cell wall material (Wilson & Ford, 1971; Wilson & Minson, 1980).

The N concentration of plant organs usually decreases in the order: inflorescences $\geq$ lamina > leaf sheath $\simeq$ stems (Henzell & Oxenham, 1964); young stem may be superior to old leaf. The recognition grew that species differences in total N% of shoots often reflect differences in the relative proportion of leaf; ontogeny is a powerful influence on these proportions and additionally N% of component organs is reduced in flowering tillers. Grass species may be chosen which are adapted to the desired conditions of soil fertility and management system used in particular farm situations.

3.2.5 Nutrient balance and soil acidity

The testing of soils for mineral deficiencies and the diagnosis of this by recognition of visual symptoms in plants and to a lesser extent by tissue testing have become well entrenched in farm practice. This has led to the balanced application of fertilizer and soil amendments such as lime to meet the requirements for optimum herbage growth; the specific requirements of ruminant animals also receive attention.

The movement to higher levels of N fertilizer usage gradually produced an appreciation that a concomitant need for higher levels of other nutrients had been generated. For example, at the X International Grassland Congress Andreyev & Savitskaya (1966) reported a study on a podzol at Moscow of *Bromus inermis*. Unfertilized swards produced 4.6 t ha^{-1} of hay, and yield was increased to 9.5 t ha^{-1} if 120 kg N ha^{-1} were applied; production increased to 14.5 t ha^{-1} if P and K were supplied. In Cuba Crespo & Cuesta (1974) reported similar positive interactions for the yield of *Digitaria decumbens*. Strongly positive interactions between N and P supply are widespread for both temperate and tropical grasses (Wilson & Haydock, 1971).

The increased demand for other nutrients is exacerbated in cut-and-remove systems, and traditionally K deficiency appears after repeated hay crops (Clement, 1971). Grazing provides a 'shower of fertility' which returns the greater part of ingested nutrients to the sward, as discussed later, and this reduces the fertilizer requirement estimated from cutting experiments. The nutrient requirement of the high N system might also be seen in the context of the greater specialized mineral demand presented by the alternative N source, the rhizobial symbiosis.

Soil acidification results from repeated high level N applications, unless lime is applied. Sulphate of ammonia has more deleterious effects in this respect than other N fertilizer sources (Mulder, 1952). In the Puerto Rican study of *P. purpureum* (Table 3.1), a Toa clay loam was initially limed to pH 7; after three years of application of 896 kg N ha^{-1} yr^{-1} as ammonium sulphate the pH decreased three units (Vicente-Chandler *et al.*, 1959). Teitzel, Gilbert & Cowan (1991) report a range of acidification values from 0.01 to 0.33 units of pH yr^{-1}. The change in soil reaction alters the availability of other nutrients; decreasing pH decreases plant access to Ca, Mg, P and Mo, increases the possibility of Mn and Al toxicity, and increases access to Cu, Zn, Fe and B. Many tropical grasses are tolerant of low pH. In Georgia, USA, Adams, White & Dawson (1967) observed that raising the pH of a Cecil sandy loam from 4.3 to 4.8 increased the yield of *Cynodon dactylon* cv. Coastal; no responses occurred above this level. Problems arise if high N grass pastures are incorporated in rotational systems involving less acid-tolerant crops.

3.2.6 Fertilizer technology

Urea is regarded as the least effective of all common N fertilizer sources (for example Henzell, 1971; Figarella, Abruna & Vicente-Chandler, 1972; Murtagh, 1975) but this may be compensated by lower price per unit N. The greater acidifying effect of ammonium sulphate needs to be set against its value as a source of S and its high rate of recovery by grasses. Ammonium nitrate and its various forms when incorporated with lime have had wide acceptance, and local factors of availability may favour sodium nitrate. The development of compound fertilizers, such as diammonium phosphate or ammonium phosphate, led to reduced unit transport and application costs, but the omission of S is often an unrecognized but widespread problem (Neller, 1952).

The mobility of soluble N sources, the opportunities for leaching, denitrification or volatilization of ammonia, the danger of excessive NO_3 accumulation and the need to synchronize N availability with plant demand have led to frequent N application in farm practice (for example, Kaltofen et al., 1966). An alternative approach is to develop slow-release N fertilizers, of which sulphur-coated urea was one of the first. Mays (1970) demonstrated that a single application of sulphur-coated urea gave the same growth pattern as three split dressings of ammonium nitrate.

Injection of anhydrous ammonia (Drysdale, 1970) was found to give more sustained responses than conventional N sources, but the process of injection is not feasible on some grassland locations, and adds to cost. Aqueous ammonia gained favour in some countries, and more recently there has been a considerable focus on the incorporation of nitrifying inhibitors which decrease N losses (Iglovikov, Kulakov & Blagoveschchensky, 1985). Recent results with wax-coated calcium carbide, which slowly releases acetylene, indicate that emissions of nitrous oxide and methane following urea application are much reduced (Smith, Freney & Mosier, 1993). Manufacturers continue to develop new N fertilizer products which deliver superior N recovery and patterns of N availability suited to particular farm systems.

3.3 Biological N fixation

3.3.1 The rhizobial–legume symbiosis

The rhizobial–legume symbiosis is the greatest source of biological N fixation. In this phenomenon atmospheric N_2 is reduced by the enzyme nitrogenase to NH_3 and H_2 and transformed to glutamate and other compounds through the link with bacteria inhabiting the root (and stem) nodules of legumes, which then acquire a source of combined N which reduces their dependence on native soil N for growth. The significance of these processes is illustrated by an affirmation at the VI International Grassland Congress by a legume missionary, J. Griffiths Davies: '. . . the results obtained have revolutionized the agricultural potential of southern Australia, for not only is the pasture potential transformed but the uplift of fertility eventually to be achieved will lead to a more intensive agricultural regime' (Davies, 1952).

Russell & Williams (1982) estimated an annual fixation in Australian pastures of 0.8 Mt N.

Such transformations depend upon environmentally adapted legumes, as discussed in chapter 2, superior rhizobia in terms of N fixation and ecological success, and the management of the host–bacterium association to minimize constraints on its effectiveness. A primary consideration is the energy cost of N fixation.

Early studies suggested that the growth of legumes was much reduced by the diversion of carbohydrate to the nodules where most estimates indicated that the ratio of carbohydrate required : N fixed fell within the range 6:1 and 20:1 (cited by Gibson, 1966). A.H. Gibson was critical of the approaches used. When he distinguished the cost of nodule establishment from N fixation and nodule maintenance and compared nodulated plants with plants assimilating the same level of N as ammonium nitrate, a similar energy cost for N fixation and for uptake of combined N was obtained (Gibson, 1966). It was subsequently realized that the reduction of a molecule of atmospheric N_2 requires a concomitant reduction of at least two protons to H_2 (Giller & Wilson, 1991, to whom some of this review is indebted), and that electrons are acquired by intermediary enzymes from a strong electron donor. Nevertheless, Herridge & Bergersen (1988) and Henzell (1988) have confirmed Gibson's view that the cost of N fixation can be safely carried by legume swards with little loss of production.

A second consideration is the degree to which legumes depend upon N fixation and upon uptake of native soil N. Many early assessments of N fixation (for example the estimate of Hutton & Bonner, 1960, of 560 kg N ha^{-1} yr^{-1}) were based simply on shoot N uptake, or upon the difference between nodulated legumes and a control treatment, without differentiation of N source. A further complication was the 'priming' effect of added available nitrogen, which increases the access to and uptake from native soil N (Jenkinson, Fox & Rayner, 1985). These difficulties have been partially overcome by the development of new techniques.

The proportion ('P') of N uptake derived from N fixation was estimated by Vallis *et al.* (1967) using small additions of ^{15}N-labelled fertilizer. *Stylosanthes humilis* grown in pots in competition with *Chloris gayana* acquired 47% of its N uptake in its first 35 d of seedling growth from soil sources, but subsequently became dependent upon rhizobial bacteria and in total derived 97% of its N from this source. This work was followed by more widely ranging field experiments (Vallis, Henzell & Evans, 1977) which showed that for a range of legumes in mixed swards some 92–94% of legume N uptake had been fixed. In cropping situations, perhaps involving fallowing and fertilizer N application, crop legumes may use little nodule N; in mixed swards grasses effectively capture most of the available soil N and herbage legumes rely on biological N fixation.

For some years the production of ethylene from acetylene was used to indicate nitrogenase activity, but there are many sources of error (Witty & Minchin, 1988). A more recent development is the analysis of legume cell sap to determine the proportion of indicator compounds; N is transported from nodules as ureides or amides, according to plant species. Many tropical legumes transport the products of N fixation as ureides, and an assay for ureides, total α-amino N and nitrate provides an estimate of 'P'; an alternative technique which estimates the natural abundance of isotopic forms of N has also gained strength (Herridge, Bergersen & Peoples, 1990).

3.3.2 Rhizobial strains and specificity

The genetic constitution of rhizobial strains determines their affinity with particular host legumes, both in terms of the capacity to form nodules and of the effectiveness of N fixation. The concept of the cross-inoculation group, in which certain host species are nodulated by one set of rhizobial strains and not by others, became well established (Fred, Baldwin & McCoy, 1932). At the VII International Grassland Congress J.M. Vincent (1956) characterized strains according to their antigenic constitution, which created serotypes (Vincent, 1942), and to their symbiotic behaviour. He drew attention to the great diversity of indigenous strains found in a locality and the adaptation between the rhizobia and the clover hosts in an area, which sometimes influenced the success of inoculation when treated seed were sown. A considerable industry for the production of elite inoculants arose which contributed to the success of legume sowings, especially in areas where the planted legumes had not previously occurred.

The emphasis in temperate herbage legumes on the specificity of the host–bacterium association and its sensitivity to low pH and low Ca availability led to a counter-reaction by D.O. Norris (1956). At this time the rhizobia nodulating tropical legumes were mainly grouped in the 'cowpea miscellany'. Norris was impressed by the predominant tropical distribution of legume species and by Tutin's (1958) emphasis on the evolution of the legume from ancestral roots in a wet tropical environment, 'essentially that of the monkey, macaw, parrot and fruit-eating bat'. Norris affirmed a distinctive tolerance of acidity for tropical pasture legumes and for their co-evolving rhizobia. He proposed that bacteria which were slow-growing and alkali-producing in culture (subsequently renamed *Bradyrhizobium*, Krieg,

1984) might inoculate seeds pelleted with rock phosphate, whilst the fast-growing, acid-producing *Rhizobium*, characteristic of the temperate legume symbiosis, might continue to be inoculated on seeds pelleted with lime (Norris, 1967). A further genus (of uncertain status), *Azorhizobium*, nodulates stems (Dreyfus, Garcia & Gillis, 1988).

The simple categorization of legume species into those with promiscuous rhizobial affinities and those with specific strain requirements has given place to a recognition of the complexity of these genetic relationships and their modification by geographical and edaphic factors (Date, Burt & Williams, 1979). This is well illustrated by a study of the genus *Stylosanthes*, which was once regarded as containing mainly species which are freely and effectively nodulated by indigenous rhizobia. The selected data (Date & Norris, 1979) refer to 138 accessions of *Stylosanthes* grown in Leonard sand-jars and inoculated with 22 rhizobial strains having features of interest. The effectiveness of N fixation was rated in five classes according to the DM of inoculated legume relative to the DM of legume grown with combined N; 0–22% was regarded as ineffective and >167% as very effective.

The general relationships of strains and effectiveness of response is shown in an analysis of principal coordinates (Figure 3.4). Vector I displays the variation associated with the number of strains effective in N fixation, whilst Vector II shows the variation associated with host specificity for effective N fixation. The 22 strains are circled with dashed lines which represent affinities derived from another classificatory technique, 'GROUPER'. The upper section of Vector I indicates the strains which are generally ineffective or are specific for a limited number of accessions. Thus, the unusual behaviour of CB2126 and CB2152 may be related to their occurrence in alkaline soils and their specificity for *S. hamata*. Conversely, strains in the lower section of Vector I are generally effective with a wide range of accessions and often originated in wetter regions of acidic soils.

The actual performance of one member of each of the six rhizobial groups shown in Figure 3.4 on selected hosts (Table 3.2) illustrates the difficulty of finding elite, effective strains for some accessions, such as *S. capitata* CPI 40238, *S. guianensis* CPI 34919, and *S. hamata* CPI 40264 A. A cultivar such as Schofield showed a high level of effectiveness across many strains whilst other accessions successfully fixed N if the appropriate marriage could be arranged. The compatibility between host and rhizobia depends upon an early recognition signal which leads to bacterial multiplication and infection – whether through root hairs, a lateral crack in a root or direct penetration between epidermal cells (Giller & Wilson, 1991). It is now known that this recognition arises through the stimulation of

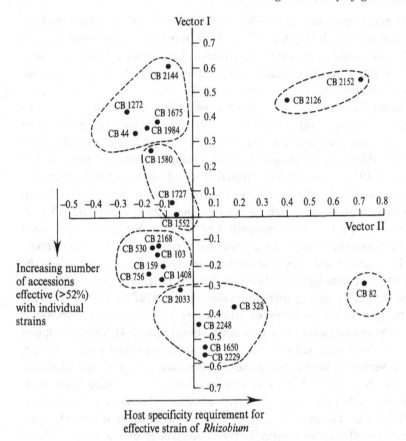

Figure 3.4. Principal coordinates analysis plot of Vectors I and II for 22 rhizobial strains used in screening effectiveness of response of 138 accessions of *Stylosanthes*. (From Date & Norris, 1979.)

biochemical activity in rhizobia by specific flavonoids in root exudate (Peters, Frost & Long, 1986).

3.3.3 Rhizobial ecology

Technologies for successful inoculation of legume seeds have been developed. These depended on cool storage of inoculants, avoidance of exposure to sunlight, use of protective and convenient media such as peat, delivery of sufficient bacteria (>3000 per seed), and use of adhesive materials (Brockwell, 1962) such as gum arabic or methyl cellulose. On acid soils temperate legumes benefited from lime pelleting (Loneragan *et al.*, 1955).

Table 3.2 *Range of nitrogen fixation response for* Stylosanthes *accessions against representative rhizobial strains*

	Rhizobial strain (CB no.)					
Accession	44	82	159	1552	1650	2126
S. capitata						
CPI 40238	I	I	I	I	I	I
S. erecta						
CPI 34118	e	e	e	e	e	e
S. fruticosa						
CPI 41219	I	E	e	ie	E	e
S. guianensis						
cv. Cook	I	e	e	ie	e	ie
cv. Endeavour	I	e	e	ie	e	ie
cv. Oxley	I	E	e	e	e	I
cv. Schofield	e	E	VE	e	E	ie
CPI 34919	I	ie	I	I	I	I
S. hamata						
cv. Verano	ie	E	ie	ie	e	e
CPI 40264	I	I	I	I	I	VE
S. scabra						
cv. Seca	I	ie	ie	I	e	e

Note:
I, ineffective; ie, poorly effective; e, moderately effective; E, effective; VE, very effective.
Source: Date & Norris (1979).

Recommendation of elite bacterial strains was initially based on nodulation and superior N fixation in laboratory tests. It became clear that an element of incursiveness was necessary to ensure field success, since the population density of indigenous bacteria in the rhizosphere of seedlings was much greater than the density inoculated with the seed, and much ineffective nodulation resulted (Date & Brockwell, 1978). Even the nodules of species regarded as having relatively specific requirements, such as *Leucaena leucocephala*, reflected the invasion of indigenous strains; applied *Rhizobium* might initially form 100% nodules, and after two years account for only 12–16% nodules (Bushby, 1982). This type of phenomenon has weakened the thrust to seed inoculation, although the contribution of the inoculant to providing vigorous, well nodulated seedlings needs to be recognized. Research activity directed to the recognition or genetic manipulation of elite rhizobia continues (Mytton & Skøt, 1993).

The soil environment influences the fate of rhizobia on inoculated seed or in soil. In hot environments soil temperatures well in excess of 55 °C occur in the top 2 cm of soil for up to 6 h d^{-1}. These are lethal to the rhizobia of both tropical and temperate species (Bowen & Kennedy, 1959); Chowdhury, Marshall & Parker (1968) found 35 °C lethal for rhizobia from *Trifolium* and *Lupinus*. These observations have led to modifications in sowing practice which ameliorate soil temperature.

Cold temperatures are inimical to rhizobial multiplication, but more significantly reduce N fixation through effects on the growth of the host. Since N fixation of *T. repens* is linearly related to temperature over the range 5–23 °C it is estimated that cool temperature limits N fixation in the UK to *c.* one half of its potential (Institute for Grassland and Animal Production, 1989). Drought and waterlogging are further factors adverse for rhizobial density in soils. The widespread occurrence of vesicular arbuscular mycorrhiza (Bethlenfalvay, Ulrich & Brown, 1985) increases the uptake of P on low P soils and these fungi interact positively to increase nodulation and the effectiveness of N fixation.

3.3.4 N fixation and mineral nutrition

Soil fertility influences the level of N fixation through effects on (i) the multiplication and survival in soil of the rhizobia which infect legume roots, (ii) the processes of N fixation in the nodule and (iii) the growth of the legume, which influences the supply of energy to the bactcroid and the transport of N products from the nodule.

Soil N

Mention was made earlier of the availability of soil N, which influences the dependence of the legume on fixation. Scientists sometimes follow false trails. Early nodulation of seedlings is desirable; when P.S. Nutman (1957) observed precocious nodulation on seedlings replanted on agar slopes he canvassed the hypothesis that root secretions from the seedlings grown earlier might have stimulated nodulation. However, Nutman had used tap water containing 6 mg l^{-1} NO$_3$-N in his experiments; Gibson & Nutman (1960) reported that the positive 'pre-planting effect' was due not to root secretion but to the first seedlings absorbing the nitrate in the medium which inhibited nodulation. The use of compound fertilizers containing N on mixed swards, and the unavailability of N-free fertilizers in some countries, reduces both legume nodulation and the competitive status of legumes.

Micronutrients.

The emphasis on mineral nutrition controlling N fixation, which was well accepted with respect to P, K and lime at the time of the VI International Grassland Congress in 1952, expanded in scope when new findings about the significance of micronutrients appeared (Davies, 1952; Riceman, 1952). Over great areas of southern and western Australia the application of single superphosphate, supplying P, S and Ca, was insufficient for the establishment of *T. subterraneum*. In some districts Cu, which is involved in the N-fixation process and the growth of the legume, and to a lesser extent Zn, which may be involved in the synthesis of leghaemoglobin, needed to be applied as copper and zinc sulphates, each at 8 kg ha^{-1}. Deficiency of Mn and B appeared under alkaline conditions.

Perhaps the most dramatic responses arose from increased Mo supply, since Mo is a constituent of the Mo–Fe protein of nitrogenase. Anderson (1956) indicated the difficulty of relying on soil analysis for the diagnosis of deficiencies, especially when soil pH determines availability; the first responses to Mo application arose on acid ironstone soils with a high content of Mo. In northern New South Wales the establishment of clover was promoted by the application of 2 t ha^{-1} of lime; band application of 200 kg ha^{-1} had a similar effect. However, in circumstances where lime was simply increasing the availability of Mo the alternative of 60 g ha^{-1} Mo substantially reduced fertilizer costs. The interpretation of fertilizer responses requires skill; an apparently straightforward response to lime and manganese sulphate was actually due to the occurrence of Mo and S deficiencies (Anderson & Arnot, 1953).

Co, which is a common deficiency of ruminant diets (Norton & Hales, 1976), has a significant role in cobalmin dependent enzymes in rhizobia (Dilworth, Robson & Chatel, 1979) but apparently not directly in host growth. Fe is a constituent of both nitrogenase proteins and of leghaemoglobin, and more recently the function of Se and Ni has attracted interest (Giller & Wilson, 1991).

Soil acidity

Soil acidity, often associated with low availability of Ca and Mg and with toxicities of Al and/or Mn, is inimical to the survival of rhizobia of clovers and medics (for example, Loneragan & Dowling, 1958). On the other hand the rhizobia of some tropical legumes require culturing in acid conditions, and their failure to multiply in a neutral medium may cause them to be overlooked in routine testing which does not recognize the diversity of response (Date & Halliday, 1979).

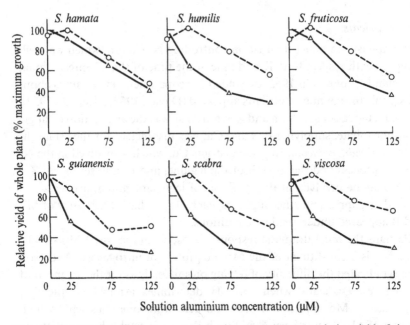

Figure 3.5. Effect of aluminium concentration on percentage relative yield of six *Stylosanthes* spp. either receiving 0.5 mM ammonium nitrate (--) or inoculated with CB756 *Bradyrhizobium* (–). (From de Carvalho, 1978.)

The rhizobial symbiosis is difficult to optimize under acid conditions linked to mineral toxicities. For example, de Carvalho *et al.* (1981) grew six species of *Stylosanthes* at Al concentrations of 0, 25, 75 or 125 μM with basal calcium at 0.4 mM. The relative yield of these treatments is shown either for plants receiving 0.5 mM ammonium nitrate or for plants dependent upon symbiotic N fixation (Figure 3.5). There are marked differences between species responses in the nodulated series; *S. hamata* and *S. fruiticosa* were superior to *S. guianensis, S. viscosa* and *S. scabra*, with *S. humilis* intermediate. Shoots were reduced more than roots in the sensitive species and Al accumulation was less in the shoots of the tolerant species. Ca and Mg uptake seemed adequate in all treatments, but K concentration was marginally lower in the sensitive species. Figure 3.5 illustrates the sensitivity of N fixation to Al toxicity; plants receiving combined N produced much greater relative yields at high Al levels. *S. hamata* cv. Verano forms a notable exception; it shows a remarkably efficient symbiotic association with *Bradyrhizobium* at the highest Al concentrations and little reduction at 25 μM.

S. hamata nodulated at all Al levels; *S. humilis* was superior to the other species, whilst *S. scabra* and *S. viscosa* formed few nodules at 75 and 125 μM. *S. hamata* had the highest ratio of nodule weight to plant weight. Both the N concentration in shoots and the N uptake of sensitive plants were greatly reduced when dependent upon the rhizobial symbiosis. The capacity of legumes not only to grow under conditions of high soil Al but also to add N to the soil–plant–animal system under these adverse circumstances is recognized in plant improvement programmes as a selection criterion of profound significance. More recently Al saturation is regarded as an ineffective index of potential plant toxicity, and this derives from the monomeric forms of Al and not the organic Al polymers (Alva *et al.*, 1986). The former reduce the absorption of other minerals and confer a rigidity to the membrane of root cells which impairs cell expansion (Blamey *et al.*, 1993).

A negative consequence of the successful adaptation of herbage legumes is that the production of nitrate eventually contributes to soil acidity. Williams (1980) noted that 50 years of *T. subterraneum* pastures at sites in New South Wales led to a reduction of one unit of pH; in poorly buffered sandy soils Jarvis & Robson (1983) observed that *T. subterraneum* affected the cation/anion balance and effected changes in soil pH. The use of deep rooted perennials is one approach towards ameliorating this problem.

Fertilizer practice

The dependence of N fixation on fertilizer inputs on soils of low nutrient status has long been demonstrated, but its acceptance by farmers has been limited by the rising costs of nutrients and the decreasing terms of trade for many agricultural products (Pulsford, 1980). The powerful interaction of P supply, N fixation and plant introduction was well illustrated by a study of *Phalaris tuberosa – T. subterraneum* swards (Anderson & McLachlan, 1951) in which production relative to an adjacent native *Danthonia* sward was increased by a factor of seven. The fertility-building phase of planted pastures was accelerated by the application of a heavy dressing of 89 kg P ha^{-1} in the first year; this led to initial clover dominance and a subsequent grass resurgence as the N supply was augmented. Low maintenance P would maintain a significant clover component in a grass-dominant sward. Thus, the economic response to P application is judged not only in terms of DM production and P concentration but also through its effects on N uptake; the range in response on mixed swards in south east Queensland was 0.2–1.9 kg additional N kg^{-1} P according to legume species and P application level (Blunt & Humphreys, 1970). The rate of soil N accumulation under a *Desmodium uncinatum* pasture was 71 and 126 kg N ha^{-1} yr^{-1} respectively under a low and high P fertilizer regime (Henzell, Fergus &

Martin, 1966). The low input approach which eschews fertilizer application may impose constraints on N fixation which preclude subsequent crop responses to a legume pasture ley (Gibson, 1987).

3.3.5 *Other organisms*

Some non-leguminous shrubs and trees, which have a place in silvopastoral systems, have N-fixing organisms in their roots. The soil actinomycetes of the genus *Frankia* form a symbiotic N-fixing association with *Casuarina* spp. which may contribute 40–60 kg N ha^{-1} yr^{-1} (Dreyfus *et al.*, 1987). This association is also more effective under conditions of adequate P supply (Reddell *et al.*, 1988).

The discovery of increased N uptake in grasses associated with the bacterium *Azospirillum* spp. (Day, Neves & Döbereiner, 1975) led to the diversion of considerable research resources into this area in the 1970s and 1980s. This has not led to any significant innovation in agricultural practice, although some attempt to commercialize inoculation has been made in Israel (Boddey & Döbereiner, 1988). The levels of additional N uptake claimed are usually in the range 0–30 kg ha^{-1} yr^{-1} and rarely result in any increase in N% of shoots. The mechanisms of response are still debated; it is known that *Azospirillum* stimulates the production of growth hormones, and growth responses to inoculation are linked to increased plant uptake of soil and fertilizer N, associated with increased root growth (Bashan, 1993). The lack of a specific structural formation on the roots makes the organism vulnerable to competition and to the leaching of beneficial products. The field effects have been unpredictable and unreliable. Earlier research attention to other free-living bacteria, such as *Azotobacter, Beijerinckia* and *Derxia*, has not been significantly extended to agricultural practice.

3.4 The N cycle under grazing

The N cycle is the largest nutrient cycle, both in terms of the size of nutrient pools and the flows which occur between them. This creates opportunity for the greatest errors in estimation, but also accords great flexibility to the fate of N as desired animal product, sequestered organic matter or environmental pollutant.

At the VI International Grassland Congress Mulder (1952) referred to the need for 'a close examination of the fate of unabsorbed fertilizer nitrogen applied on grassland', as recalled by Henzell (1970) when seeking better progress in understanding with precision the component processes and their significance in the N cycle. However, by this stage drainage lysimeter studies had been long in progress, ^{15}N was used in measuring N fluxes and gas lysimeter techniques were being developed. Henzell's review referred to the 'risks to health of man and animals from high levels of nitrate in drinking water' and by the time of the XIV Congress in 1981 the research focus on efficient production had been balanced by increased attention to N sources as environmental pollutants (Lazenby, 1983).

3.4.1 Case studies

Two systems of contrasting intensity and product output are used to introduce the topic.

Beef production in central Brazil

A mixed *Brachiaria decumbens* sward containing *c.* 20% of the relatively unpalatable legume *Calopogonium mucunoides* growing on a red latosol at Campo Grande, Brazil (lat. 23° S) was grazed by cattle at 2.2–2.5 AU ha^{-1} (AU, animal unit = 450 kg) (Cadisch, Schunke & Giller, 1994). The N pools and fluxes (Figure 3.6) were estimated from pasture and soil sampling, LWG and extrapolation from other studies. The system was regarded as being marginally in positive balance with respect to N, whilst an adjacent pure grass sward was estimated to be losing 65 kg N ha^{-1} yr^{-1} from the soil OM reserve. Research elsewhere suggests the rundown of grass-only pastures may also be associated with immobilization of N in litter (Robbins, Bushell & McKeon, 1989).

The plant biomass (Cadisch *et al.*, 1994) had an internal N cycling component and 90% of the legume N shoot uptake of 93 kg N ha^{-1} yr^{-1} was regarded as N fixation. Green grass averaged 1.06% N and legume 2.17% N. The litter component was of a similar order to the animal N intake, which led to 535 kg LWG ha^{-1} yr^{-1} and a product output of 13 kg N ha^{-1} yr^{-1}. The latter figure represents a very small proportion of the N being circulated; it might be noted that successful efforts to increase the diversion of N to animal product do not necessarily reduce environmental pollution, since this eventually requires increased disposal of human waste N. The model postulates a loss of 50% N in animal excreta, whilst the balance goes to microbial biomass and soil OM.

The soil OM to 1 m depth constituted a large reserve of 7.5 t N ha^{-1}, of which only a small fraction would be active in the N cycle. Net mineralization was estimated as 157 kg N ha^{-1} yr^{-1}, and deposition of N from dust and rain was hopefully balanced by a low estimate of leaching and denitrifica-

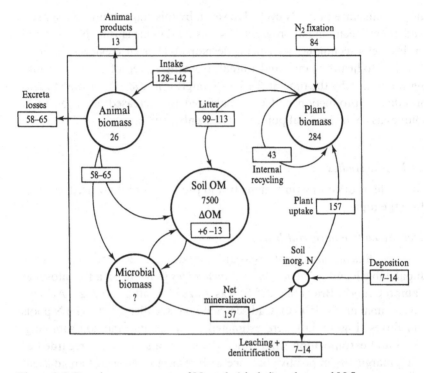

Figure 3.6. Putative components of N pools (circled) and annual N fluxes (rectangles) as kg N ha⁻¹ in a *Brachiaria decumbens – Calopogonium mucunoides* sward in Campo Grande, Brazil. (From Cadisch, Schunke & Giller, 1994.)

tion from soil inorganic N. A model with similar overall numbers was developed for a grass–legume pasture in the llanos of Colombia (Thomas, 1992). These studies forsee the introduction of a legume as reversing current trends of N drainage from the pure grass swards which dominate the cerrados and llanos grasslands of South America.

Dairy production in south west England

More precise estimates of N losses are available from the intensive research carried out on UK and other European pastures. Jarvis (1993) describes a model dairy farm in south west England comprising 25 ha of aged sward on poorly drained clay/loam soils and 51 ha of reseeded sward on moderately drained loam. The average herd comprised 102 cows each yielding 5550 l milk, and 110 other cattle giving a total of 165 AU. It is assumed that 129 t concentrate are purchased and 394 t silage DM are made.

The base system had inputs of 250 kg fertilizer N ha⁻¹ yr⁻¹, 10 kg N from fixation, an optimistically stated 25 kg from atmospheric deposition, and

Table 3.3 *Annual losses of N (kg ha⁻¹) and support energy (GJ ha⁻¹) required for differing production systems on a model dairy farm in south west England*

Source	Base system	Tactical N with slurry	Grass– clover	Grass–clover and maize silage
N losses through grazing:				
Leached	34	24	18	13
Denitrified	40	27	21	17
Volatilized	10	6	5	4
N losses from wastes:				
Leached	12	12	9	2
Denitrified	11	11	9	2
Volatilized	35	16	29	14
Total N losses	142	96	90	54
Loss of N_2O-N	13	10	7	5
Support energy	29	26	7	7

Source: Wilkins (1993).

52 kg ha⁻¹ average from purchased feeds, totalling 337 kg N ha⁻¹. Additionally, soil mineralization was estimated as 157 kg N ha⁻¹.

The output of milk accounted for only 39 kg N ha⁻¹, and if a further 28 kg N assimilated into the protein of the younger animals is added, the equivalent of only 27% of the fertilizer N input is represented. The losses in the system are summarized by Wilkins (1993) (Table 3.3), who compared other options with the base system. It was assumed from other studies that 78% of consumed N was excreted. Losses from leaching, denitrification and volatilization were of similar magnitude when the direct losses from grazing were added to the losses from housed animals and from spreading slurry, and these totalled 142 kg N ha⁻¹, indicating a large unaccounted component, much of which must represent accretion to soil OM.

The input of N fertilizer may be reduced without loss of herbage production if tactical application is made to avoid excess accumulation of mineral N. A rapid test (Titchen and Scholefield, 1992) can be made to ensure that fertilizer N is only used to rectify deficiencies in supply in relation to grass requirements, and to take account of N mineralization and returned slurry and excreta. If N fertilizers were reduced to 200 kg N and slurry injected to decrease ammonia loss N pollution (Table 3.3) would be appreciably reduced.

The alternative of zero fertilizer N and the promotion of white clover/grass swards might require a reduction in SR to 80%, but might be expected to decrease NO_3 leaching. A further option would be to grow maize to substitute for half of the herbage silage, and to plough in slurry (which would also be reduced in amount because of the more efficient digestion of maize silage); this option, in combination with clover-based grazing, would cut total N losses to 54 kg N ha^{-1} yr^{-1}.

The partitioning of denitrification to identify the greenhouse gas N_2O indicates that the last two options effect reductions; the support energy required in grass and forage energy would also decrease.

3.4.2 Transfer of legume N to grass

The phenomenon of transfer of legume N to companion grass is indubitable, but the pathways and mechanisms of transfer (which were recently modelled by Thornley, Bergelson & Parsons, 1995) are much disputed. Earlier mention was made of N excretion from nodules (Virtanen, 1937) and the discounting of this process by Butler & Bathurst (1956). At the VII International Grassland Congress these authors drew attention to the senescence and sloughing of legume nodules, and estimated that in New Zealand 81 kg N ha^{-1} yr^{-1} might be released to the soil from a *T. repens* sward by this process. The effect of defoliation on premature senescence is discussed in the next chapter. Although nodules have a high N%, it is unusual for them to comprise much of the N pool at any stage; for example, the nodules of undefoliated *Desmodium uncinatum* contained 1–5% of total plant N content (Gibson & Humphreys, 1973). Turnover of root tissue is a further potential mechanism of N transfer, but usually root N constitutes less than one third of the plant N pool, depending upon species and management, although Sanginga *et al.* (1990) refer to an experiment with *Leucaena leucocephala* where about 60% of total plant N was located in the roots. However, the size of the N pool is less important than the N flux; thus, heavy grazing reduces root growth but increases the instantaneous root:shoot ratio. Many cut-and-remove systems of use result in little N transfer to grass (for example, Catchpoole & Blair, 1990).

The majority of scientific opinion emphasizes the transfer of legume N by animal excreta (Walker, Orchiston & Adams, 1954) and by the decomposition of legume leaf and stem litter (Whitney & Green, 1969). In Figure 3.6 the litter component comprises more than a third of the N flux. The proportion of N cycled through the animal rather than the litter pathway depends upon SR, and the gaseous and leaching losses of N from excreta under high SR are partly ameliorated by the faster rate of N cycling and the lower C:N ratio of the system (Mears & Humphreys, 1974a). Conversely, it may be argued (Steele & Vallis, 1988) that plants with a high C:N ratio reduce N losses since more N is incorporated into relatively stable OM.

Legume leaf litter is degraded to OM at varying rates, according to environmental conditions and species characteristics such as polyphenol content (Vallis & Jones, 1973; Gutteridge, 1992). The half life for N disappearance was 22 and 53 d respectively for litter of *Gliricidia sepium* and *Flemingia macrophylla* (Budelman, 1988). The problem in agricultural practice of attaining synchrony between legume N release and grass N demand may be partly addressed through appropriate choice or admixture of legume species.

3.4.3 Leaching losses

Loss of NO_3 through leaching or through surface runoff reflects an inefficiency in the production system; it has also become of considerable moment for the pollution of rivers and aquifers. The delivery of slurry or other excretory wastes or silage effluent continually cause point pollution and increase the biological oxygen demand, leading to eutrophication. These events occur more frequently in catchments of high SR (Wilkins, 1993). The maximum permissible level of NO_3 in the European community has been set at 50 mg NO_3 dm^{-3} (= 11.3 mg NO_3-N dm^{-3}) and agricultural practice is directed to minimizing infringements of this standard, which frequently arise in intensively farmed areas.

Initially, clover based swards were regarded as environmentally 'clean', but the early comparisons were based on cut swards where leaching was greater from areas receiving high levels of fertilizer N. If swards were grazed the N recycled in excreta led to higher levels of leaching than occurred in cut swards (Ryden, Ball & Garwood, 1984), and soil NO_3-N behaves similarly irrespective of source.

A classic study at North Wyke, south west England, illustrates some of the influential factors (Scholefield *et al.*, 1993). Long established pastures were fertilized at 200 or 400 kg N ha^{-1} yr^{-1} and pastures reseeded to *L. perenne* were fertilized at 400 kg N ha^{-1} yr^{-1}. Nitrate leaching was measured from drained and undrained plots. Loss of NO_3 (Figure 3.7) reached dangerous levels on old pastures receiving 400 N, and was less at 200 N; it averaged respectively 134 and 39 kg NO_3-N ha^{-1} yr^{-1}. There would have been a great quantity of NO_3 lost from the reseeded sward in 1982–3 after ploughing, but subsequently leaching decreased relative to leaching from old pastures, as depicted in Fig. 3.7.

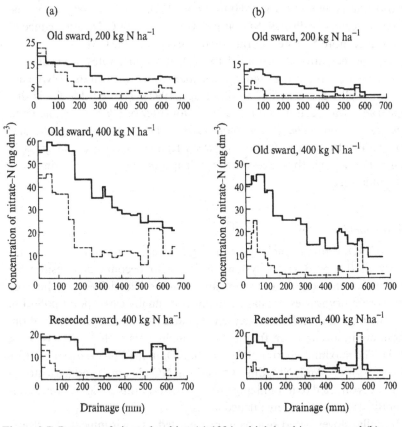

Figure 3.7. Patterns of nitrate leaching (a) 1984, a high leaching year; and (b) 1986, a low leaching year, for different swards on drained (–) and undrained (--) plots. (From Scholefield *et al.*, 1993.)

Leaching was greater by a factor of two in 1984 after a hot dry summer than in 1986 after a cool, wet summer. This illustrates the difficulty of predicting safe levels of N application, since this is season and site specific, but points to the advantage of reducing fertilizer N level in dry seasons. Installation of drainage increased leaching losses. The interdependence of leaching and denitrification (aerobic conditions, high leaching and little denitrification; anaerobic conditions, low leaching and much denitrification) was not in balance; drainage led to a reduction of 30% in denitrification but *c.* 300% increase in NO_3 leaching. Jarvis, Scholefield & Pain (1995) suggest that for farm drainage water to meet EC NO_3 standards annual loss should not exceed *c.* 30 kg N ha^{-1} on a drained, well-structured clay loam and *c.* 20 kg N ha^{-1} on a poorly drained soil of similar texture.

Nitrate leaching is minimized when soil mineralization and the addition of N sources are in balance with sward demand. Reference was made to tactical N application based on soil testing (Titchen & Scholefield, 1992); plant NO_3-N is another indicator (Wilkinson & Frere, 1993). Reduction of the grazing season in the autumn is not an appropriate approach, since although plant N demand is reduced in the autumn the positive association of mineralization and soil temperature restricts the amount of mineralization. The avoidance of slurry application to grassland during the cold season reduces leaching. The addition of deep rooted herbs such as chicory have been advocated as a means of increasing efficiency of N use (Ruz-Jerez, Ball & White, 1993).

3.4.4 Gaseous losses

Denitrification and volatilization of ammonia, as illustrated in Table 3.3, contribute in different ways to atmospheric pollution and to productivity loss. This area of grassland science is bedevilled by problems of technique and by the complex interactions of many factors.

Denitrification

This is primarily a biological process dependent on microbial action in which there is a progressive reduction of NO_3^- acting as a terminal electron acceptor in bacterial respiration under anoxic conditions to release N_2O and N_2 (Jarvis *et al.*, 1995). The process depends upon the supply of NO_3 and is greater at high levels of N supply (Jarvis, Barraclough, Williams & Rook, 1991); it is especially evident if fertilizer application is made when soils are wet or immediately preceding rainfall. The level of ammonium, which leads to the production of NO_3, has also been implicated and the suggestion is made that the processes of nitrification and denitrification may be coupled (Jarvis *et al.*, 1994).

The level of N_2O emission is critical, since this gas has a high radiative forcing action in the atmosphere and may also be implicated in ozone depletion. The normal ratio of $N_2O:N_2$ of 1:3 may be increased substantially as the rate of denitrification increases. Temperature exerts a primary and positive control of denitrification, which is therefore of little moment in autumn and winter, despite the frequent occurrence of wet anoxic conditions.

A further factor is the supply of labile C substrate, which is usually readily available in grassland. Denitrification is increased by the application of slurry, which supplies both a soluble C and an N source, as demonstrated

Table 3.4 *Faecal and urinary outputs from ewes grazing different swards and ammonia losses at Hurley, UK*

Component	Grass	Clover	Grass– clover	Grass 420N
		Sward type		
Faecal output (g DM ewe^{-1} d^{-1})	240	240	260	270
N%	3.0	3.9	3.1	3.5
Faecal N output (g N ewe^{-1} d^{-1})	7.3	9.4	8.0	9.7
Urinary output (g fresh wt ewe^{-1} d^{-1})	3500	5200	3530	4290
Urinary N output (g N ewe^{-1} d^{-1})	32.4	55.0	36.0	53.6
Urinary N (% total N)	81	85	82	85
N-excreted (g ewe^{-1} d^{-1})	39.7	64.4	44.0	63.3
Stocking rate (ewes ha^{-1})	19	27	27	37
Total N excreted (kg N ha^{-1})	160	360	250	480
Ammonia loss (kg N ha^{-1})	4	11	2	10

Source: Orr *et al.* (1995); Jarvis, Hatch, Orr & Reynolds (1991).

at Zegveld, Netherlands by Jarvis *et al.* (1994). However, in the Basque country of Spain Estavillo, Gonzalez-Murua & Rodriguez (1993) estimated less denitrification following the application of cow slurry than following the application of equivalent amounts of N as ammonium nitrate. The time sequence of denitrification following the application of N sources may extend to subsequent seasons (Jarvis, Barraclough, Williams & Rook 1991).

Volatilization of ammonia

Volatilization of NH_3 occurs principally from the hydrolysis of urea by the enzyme urease. Urea is the main source of animal N excretion, but is also a significant fertilizer in many countries. Ammonium sulphate and ammonium nitrate are also NH_3 sources, McKenzie & Tainton (1993) reporting

4% of applied N as NH_3 loss. This process is favoured on alkaline soils of low CEC (cation exchange capacity) (Jarvis *et al.*, 1995). Table 3.3 indicates that in the UK utilized NH_3 loss from housed animals and from the application of slurry cause greater losses than from cows grazing.

Ammonia losses from grazing ewes are illustrated by a study at Hurley, UK (Orr *et al.*, 1995), in which excretory N was monitored from swards of grass (*L. perenne* without N fertilizer), clover (*T. repens*), grass/clover, and grass receiving 420 kg N ha^{-1}. The N output from ewes over a single grazing season (Table 3.4) indicated the predominant partition of N to urine and the greater N output of ewes grazing clover swards or grass swards receiving 420 kg N ha^{-1}. When this output was adjusted for SR the N output varied from 160 to 480 kg N ha^{-1} according to sward system.

The ammonia emissions were monitored by a mass balance micro-meteorological method (Jarvis, Hatch, Orr & Reynolds, 1991) which sampled the air above the pastures. The losses were greater during the middle of the grazing season but in these UK conditions NH_3 emission under grazing was a minor source of loss (Table 3.4), averaging 1–3% of total N excreted, according to treatment. This result might be contrasted with estimates from the dry tropics where total N losses were 46% of the urea N applied during the dry season at Katherine, Australia (Vallis *et al.*, 1985), and 16–32% at Townsville, Queensland (Vallis & Gardener, 1984).

This review has indicated the differing importance of the pathways of N cycling according to local conditions, and the great opportunities for managers to increase the efficiency of N utilization and to improve environmental quality.

3.5 Systems involving legumes and/or N fertilizer

3.5.1 Systems based on N fertilizer

The highest levels of animal production per unit area in humid environments are only attainable through high levels of fertilizer N. Production of milk or LWG per head is usually greater from legume based pastures, albeit at lower SR. This generalization has now been challenged for the subtropics and tropics by performance data on N fertilized *Pennisetum purpureum* cv. Mott in Florida, where LWG on this leafy, non-flowering C_4 grass was sustained at 1.0 kg LWG hd^{-1} (Sollenberger & Jones, 1989). One of the record claims for N fertilized grass arose from studies of irrigated *Digitaria decumbens* at Parada, Queensland (lat. 17° S). Ammonium nitrate was applied at 670 kg N ha^{-1} yr^{-1}, and over four years production averaged 2.76

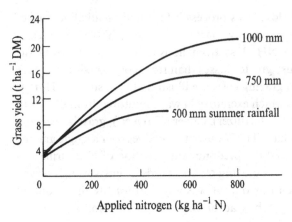

Figure 3.8. Annual pasture grass yield in relation to summer rainfall and N fertilizer rate in Queensland. (From Teitzel, Gilbert & Cowan, 1991.)

t LWG ha^{-1} yr^{-1} at 12 b (beasts) ha^{-1}, which represents 4.6 kg LWG kg^{-1} N applied (Eyles, Cameron & Hacker, 1985). This production level is more than double most commercial claims under rain-grown conditions (Teitzel *et al.*, 1991).

The moderating influence of rainfall on N response, which was illustrated for spring conditions in Figure 3.2, has been modelled for summer rainfall from cutting experiments in Figure 3.8. This may be further modified by applying a predictor of rainfall, the Southern Oscillation Index, to the expectations for N response; Gilbert & Clarkson (1993) also give an exponential equation for LWG response to fertilizer N in Queensland which is based on sites receiving more than 1125 mm annual rainfall. Near maximal LWG ha^{-1} occurred at 300 kg N ha^{-1} yr^{-1}, a fertilizer level greatly reduced under grazing relative to cutting.

On temperate pastures rates of individual LWG are higher than in the tropics. In south west England 1140 kg LWG ha^{-1} yr^{-1} was attainable on *L. perenne* at 400 kg N ha^{-1}; this represented 2.9 kg LWG kg^{-1} N applied, or 1.4 kg additional LWG kg^{-1} N applied (Wilkins *et al.*, 1987). The use of fertilizer N on grass has become well entrenched in dairy practice in Europe, North America and Japan. The assessment of milk production attributable to grass is complicated by the purchase of concentrates and the level of forage conservation. Table 3.3 was based on *L. perenne* pastures receiving 250 kg N ha^{-1} yr^{-1} and cows yielding 5550 l milk yr^{-1} when supplemented with *c.* 1.4 t concentrate hd^{-1}. Earlier recommendations (Thomas & Young, 1982) suggest target application values of 300–450 kg fertilizer N ha^{-1} according to site class, which should yield 8–12.5 t DM ha^{-1} yr^{-1} at a grazing stage (68 + D, digestibility). In Queensland high N experimental pastures have yielded 12–23 t milk ha^{-1} yr^{-1} (Chopping *et al.*, 1976, 1982).

The pressure to move to less intensive systems of production arises from two sources. On the one hand societies are becoming less willing to subsidize agricultural production; intensive agricultural systems operate nearer to the margin for benefit:cost ratios and currently are only surviving in communities which artificially channel the flow of economic resources for their support. The price farmers received for their milk (Mitchell, 1993) was for New Zealanders 46% of the price US farmers received, and respectively 35% relative to the Dutch, 33% to the Germans, and 19% to the Japanese. The liberalization of trade policies is gradually moving farmers away from high N fertilizer systems.

The second pressure arises from environmental concerns. The primary energy cost of burning fossil fuel to manufacture N fertilizer compares adversely with the biological N fixation option (Table 3.3). Whilst legume based pastures contribute to atmospheric pollution via denitrification and to stream pollution, the European experience indicates that farmer use of N fertilizer has been a greater source of delinquency.

3.5.2 Systems based on legumes

Some countries have successfully based production on legume N. At the VII International Grassland Congress Walker (1956) stated: 'Under New Zealand conditions there is little doubt that proper pasture management and close attention to the nutrition of suitable legumes is superior to dependence on combined nitrogen'. In distinguishing fertility building and fertility stable phases, T.W. Walker instanced their experience where total soil N might be raised from 0.2 to 0.8% over 20 years, representing an input of *c.* 220 kg N ha^{-1} yr^{-1}.

There have always been sceptics. Williams (1967) regarded the utility of tropical legumes as limited to systems functioning with up to 60 kg N fixation ha^{-1} yr^{-1}; Boonman (1981) stated '. . . no legume has been found to associate and persist well in combination with sown tropical grasses'. The experience of farmers and scientists is diverse. In tropical regions there are many estimates for subhumid areas receiving less than 1000 mm rainfall of N fixation in the range 30–100 kg N ha^{-1} yr^{-1}, and of 100–300 kg N ha^{-1} yr^{-1} in more humid areas (Humphreys, 1994).

For humid temperate pastures in the UK Frame & Newbould (1984) suggest swards containing 20–50% clover contribute the equivalent of 150–280 kg fertilizer N ha^{-1} yr^{-1}, whilst Hopkins, Davies & Doyle (1994) estimate rates of N fixation from swards typically containing 25% clover as 74–280 kg N in lowland swards and 100–150 kg N ha^{-1} yr^{-1} in upland swards. Table 3.3 used an estimate of 80% dairy production from grass/clover relative to grass with 250 kg N ha^{-1} yr^{-1}. One of the highest claims for LWG from the grass/legume association comes from irrigated *Leucaena leucocephala – Digitaria decumbens* at Kununurra, Western Australia (lat. 16° S). The individual performance of cattle is reduced by the very hot, humid conditions experienced, but annual LWG was 1.7 t ha^{-1} at SR of 6–7 b ha^{-1} (Petty, Croot & Triglone, 1994).

3.5.3 Integrated systems

Attempts to incorporate the virtues of legumes with respect to nutritive value and N fixation in mixed swards receiving fertilizer N have failed miserably. Lowe (1966) reported a linear decrease in the percentage of clover with increasing N application rate; Green & Cowling (1960) noted responses of <10 kg DM kg N^{-1} from grass/clover swards and responses of 30 kg DM kg N^{-1} for grass swards. The worst situation arises from moderate N fertilizer application (c. 100 kg N ha^{-1} yr^{-1}) to mixed swards; this effectively substitutes for N fixation without increasing production.

On the other hand, there are often advantages in complementing legume based pastures with grass + N swards on other parts of the farm, according to site factors and to seasonal deficiencies in the continuity of forage supply. Teitzel *et al.* (1991) describe a system in which 25% of the grazing area is reserved for high N grass with a higher cool season grazing capacity which complements the balance of the property planted to grass/legume pastures; production increased from 550 to 650 kg LWG ha^{-1} yr^{-1}.

Many grazing systems integrate the cultivation of fodder crops with permanent pastures, and the use of legume fodder banks grazed in conjunction with lower quality pastures has gained credibility (see for example, Partridge & Ranacou, 1974).

3.6 Conclusion

The ways in which the N economy of grasslands is viewed have evolved in response to social pressures leading to reduced agricultural subsidies and a lesser focus on intensive production, as the quality of the environment has gained dominant attention. The control of pollution is requiring reduced and tactical application of fertilizer N, and the development of fertilizer products which deliver N in greater synchrony with grass demand.

Research attention to the pathways and motor agencies of N cycling is needed if more efficient N use and the avoidance of wastage is to occur; especially should this be directed to gaseous N losses and to the behaviour of soil micro- and macroflora.

The greater success of biological N fixation in farm practice hinges primarily on continued advance in legume improvement, as discussed in chapter 2. It is the availability of legumes well adapted to local farm environments which provide the energy base of CO_2 fixation which primarily controls N fixation, as modified by mineral inputs. The building of soil fertility and soil quality (Lal & Miller, 1993) remains a significant challenge for scientists working on the degraded land surfaces which are prevalent over much of the globe. The development of super-nodulating plants has been a promising research development, and there are also attempts at molecular transformations which would facilitate N fixation in grasses. The question might be posed as to whether a maize plant fixing N is required, or a nodulated legume behaving like a maize plant (Paul, 1988).

E.A. Paul continued . . . 'Molecular biology together with automated instrumentation should supply a large number of the needed research tools. These tools together with the expanded application of ecological principles and mathematical modelling applied to systems agriculture should provide the background for the research necessary to achieve the objectives set out . . . earlier . . .'.

4
Growth and defoliation

4.1 Introduction

The improvement of grassland productivity depends upon the recognition of elite plant germplasm which in local environments optimizes the conversion of carbon dioxide, water and minerals to digestible herbage acceptable to livestock and which protects environmental resources from degradation (chapter 2). The definition and rectification of deficiencies of mineral nutrition and the effective management of biological N fixation (chapter 3) constitute the second line of approach, which includes the definition of genotypes adapted to the soil conditions as modified. The third strand is the manipulation of the leaf surface by management of defoliation which optimizes the sustained harvesting of herbage nutrients and maintains protective cover of the soil. This depends upon insights from plant physiology, which describes and quantifies processes occurring in plants. These insights are used in meeting other objectives of grassland improvement, and managing the defoliation of grassland is selected as the most significant physiological theme and one which has excited controversy in the past four decades.

4.2 Non-structural carbohydrate 'reserves'

Earlier phases of grassland science, for example at the IV International Grassland Congress in 1937, emphasized two aspects of plant response to defoliation: (1) the significance of total non-structural carbohydrate (TNC) in the roots and crown of the plant in controlling persistence and the rate of recovery growth after defoliation; and (2) the effects of frequency and height of cutting on pasture yield.

L.F. Graber at the University of Wisconsin was an early exponent (Graber *et al.*, 1927) of the controlling influence of labile carbohydrate in the roots and crown on the regrowth after defoliation and ultimately the persistence of *Medicago sativa*. W.A. Leukel (1937) stated that: 'Early new top growth on pastures is always produced at the expense of previously stored organic food in the lower organs of the plants . . . If the top growth of the plants is removed before the lower organs are well stored with organic foods, the next growth must again be produced at the expense of the remaining reserves . . . . Repeated removal of top growth . . . depletes the plant of its reserve foods and new top growth production generally ceases'.

Cyclic fluctuations in the levels of some carbohydrate compounds are observed in response to the frequency, intensity and timing of defoliation, climatic factors, developmental stage and N nutrition. Emphasis has been placed on the more readily mobilizable carbohydrates such as the sugars, fructosans, dextrins and starch. The concentration of pentosans, hemicellulose, cellulose and lignin are not subject to the same rapid fluctuations (Sullivan & Sprague, 1943) and these structural components were discounted as potential sources of materials which might be used in tissue repair and maintenance, and in transport to and reconstitution at meristems. Fructosans and sugars in cool, temperate climates and starch and sugars in warm climates constituted the main groups of interest (Weinmann, 1961). In both herbage grasses and herbage legumes defoliation resulted in a reduction in the content of TNC in stem bases and roots, followed by a restoration to previous levels as growth proceeded. Frequent or severe defoliation, either by grazing (Weinmann, 1948) or by cutting (Weinmann, 1949) had these effects.

The focus of seasonal changes in tropical grasses was on the depletion of TNC in winter and spring and the enhancement of TNC in the late summer–autumn (Barnes & Hava, 1963); the flowering period was regarded as especially vulnerable to defoliation (Scott, 1956). The application of N fertilizer increased the allocation of assimilate to shoot (rather than to roots) and decreased TNC concentration and amount in the roots (Adegbola & McKell, 1966). In Zimbabwe Barnes (1961) showed that a combination of fertilizer N and frequent or mid-season defoliation impaired the subsequent spring growth of *Panicum maximum*; in the colder climate of Germany Bommer (1966) found that reduced TNC level was compensated by better root growth in the spring if availability of soil N was raised. High temperatures and shade reduced the plant reserves of temperate grasses and the subsequent rate of regrowth after defoliation (Alberda, 1966).

Weinmann (1952) emphasized that depletion of TNC owing to excessive defoliation affected the capacity of the root system to absorb water and

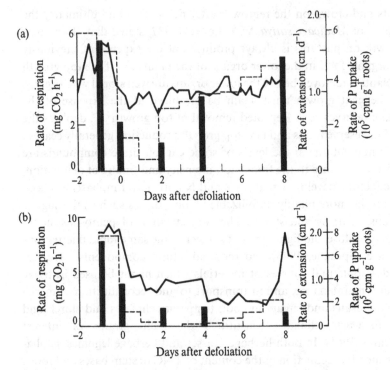

Figure 4.1. Root respiration (continuous line), root extension (broken line) and relative uptake of ^{32}P g^{-1} root weight (histogram) before and after (a) removal of laminae only and (b) cutting of *Dactylis glomerata* to a height of 2.5 cm. (From Davidson & Milthorpe, 1966b.)

nutrients. In an influential paper Crider (1955) drew attention to the cessation of root growth in several grasses which followed severe defoliation and which might persist for many days. This is elegantly illustrated (Figure 4.1) by Davidson & Milthorpe's (1966b) study of young plants of *Dactylis glomerata*. When plants were cut to 2.5 cm height (Figure 4.1b), which removed all expanded laminae and portions of leaf sheaths and expanded leaves, root respiration decreased markedly for 7 days, root extension ceased and only showed partial recovery 8 days after defoliation, whilst ^{32}P uptake was still greatly diminished 8 days after defoliation. Removal of exposed lamina (Figure 4.1a) also affected these processes but effects were less severe than in the cutting treatment; uptake of ^{32}P was more closely allied with the rate of root extension than the rate of respiration. Clearly the rate of shoot growth after defoliation might well be influenced by reduced root activity restricting nutrient uptake.

All these considerations led many grassland scientists to direct grazing and cutting management to systems which maintained a high level of TNC in the roots and crown, believing that herbage yields and plant persistence would be favoured. Weinmann (1952) suggested: 'The aim of pasture management must be to maintain an adequate level of reserves in the desirable species of the sward, which can be achieved by suitable systems of deferred and rotational grazing . . .'. The latter topic is addressed in chapter 6, and the general issue of the relevance of TNC reserves to plant response to defoliation is addressed now. Whilst there is a consensus amongst grassland scientists that TNC is involved in plant persistence and may be a factor in determining the rate of regrowth in the first 0–4 days after severe defoliation, the concept is flawed in a number of ways.

The teleological trap

A view may develop that the persistent occurrence of a plant process indicates that it is purposeful in terms of ecological success. TNC reserves were therefore regarded as an indication of plant health. This view was challenged when, in an analogous situation, Archbold (1945) suggested that accumulated sugars and fructosans in the stem of the barley plant contributed little to subsequent grain development. She proposed that stored sugar was not a purposive reserve; '. . . sugar concentration of itself plays no part in initiating plant synthesis but rather . . . stable sugar arises when assimilatory capacity is in excess of that for utilisation'. This led other workers, including Blaser, Brown & Bryant (1966), to suggest that high levels of carbohydrate in pasture swards indicate that yield potential is not being realized. TNC accumulates in tissue when growth demands are low, perhaps due to shortage of nutrients, mild water stress, or cool temperatures which constrain shoot growth. Further, it has been suggested that photosynthesis may itself be reduced by high TNC through feed-back mechanisms (Humphries, 1963). High TNC may therefore be even regarded as indicative of plant ill-health and restricted growth capacity.

Efficiency of energy use

Accumulation of carbohydrate below ground represents a use of energy which is inefficient for shoot growth, unless this is greatly reduced by the inadequacy of a root system to acquire water and nutrients. Plant material below the soil surface is inaccessible to the grazing animal, and a large mass of material with a high C:N ratio immobilizes nutrients in the bodies of organisms which form the decomposer industry, often to the detriment of shoot growth (Robbins, Bushell & Butler, 1987). The further question is the contribution of TNC reserves to shoot growth. Many (but not all) workers have found a simple correlation between TNC and the amount of etiolated growth produced in the dark (Burton & Jackson, 1962; Adegbola, 1966). Richards & Caldwell (1985) noted that dark regrowth of *Agropyron desertorum* and *A. spicatum* represented a low proportion of the carbon lost from the plant crown, and ranged from 1 to 11% of regrowth made in the light. They suggested that the contribution of carbon from reserves exceeded photosynthetically produced carbon for only 2.5 days following severe defoliation.

Plant compounds involved

Since the amount of dark regrowth following defoliation is greater than the decrease in TNC from the crown and roots over the same period (Humphreys & Robinson, 1966) and since changes in labile carbohydrate cannot account for tissue maintenance and the early regrowth of plants (Davies, 1965; Davidson & Milthorpe, 1966b; Sheard, 1970; Muldoon & Pearson, 1979b) compounds other than TNC are involved in plant recovery. Organic acids and lower order structural components such as hemicellulose may be involved under conditions of severe starvation, but nitrogenous compounds constitute a more readily available source (Culvenor, Davidson & Simpson, 1989a). Ourry, Boucaud & Salette (1988) showed with ^{15}N studies of the regrowth of *Lolium perenne* that during the first 6 days nearly all the nitrogen of new leaves arose from organic N remobilized from the roots and stubble; the free amino-acid pool was more significant than the protein or nitrate pools.

Respiratory demand

May & Davidson (1958) recorded a greater decrease in the TNC of residual tops of defoliated *Trifolium subterraneum* than in the TNC of roots, and heretically suggested that plant respiration could account for a substantial part of the decrease, and further, that translocation of TNC from tops to roots occurred. In this scenario, roots were far from constituting a source of labile carbohydrate for reassembly at the shoot meristems, and constituted a net sink. Even *M. sativa*, a plant with a long history of scientific focus on the need for the maintenance of its root reserves, was regarded by Hodgkinson (1970) as having the translocation of TNC from its roots to its shoots accounted for mainly as respiratory substrate, whilst Hendershot & Volenc (1989) also considered that root TNC of *M. sativa* may not constitute a major source of energy for the regrowth of shoots.

The carbon balance of swards of *T. subterraneum* in the first 5 days after defoliation is illustrated (Table 4.1) for swards in which 70, 30 or 0% of

Table 4.1 *Carbon utilization by* T. subterraneum *during first 5 days after 70%, 30%, or nil shoot removal as percentage of net photosynthesis of the nil shoot removal sward*

Component	Percentage defoliation		
	70	30	nil
Net photosynthesis	38	69	100
Respiration			
Shoot (dark)	4	9	14
Root + nodule growth and maintenance	11	14	14
Nitrogenase linked	4	6	11
Total	19	29	39
Dry weight change			
Shoot	16	39	53
Root	3	1	8
Total	19	40	61

Source: Culvenor *et al.* (1989b).

shoot material was removed (Culvenor *et al.*, 1989b); net photosynthesis was defined as net carbon fixed by shoots during the light period minus dark respiration for the light period. Defoliation greatly reduced net photosynthesis, according to the amount of residual leaf, and photosynthesis was directed to lamina production in the defoliated treatments. Defoliation was followed by an immediate decrease in the rate of shoot dark respiration, both as a result of removal of shoot mass and also by a reduction in the respiration rate per unit dry weight, whilst root and nodule respiration (as additionally indicated in Figure 4.1) also decreased. Respiration linked to nitrogenase activity was estimated, and decreased proportionately more than the other components of nodulated root respiration after defoliation. These values were restored to those of undefoliated swards by the 12th day after defoliation as the leaf surface was reconstructed. Table 4.1 illustrates that despite reduced respiration following defoliation the respiratory demand still constituted a high proportion of net photosynthesis.

Carbon allocation to shoots or to roots

Plant resistance to defoliation has been shown for many plants to be linked to the rapid reestablishment of photosynthetic capacity through expansion of the leaf surface. This results from many linked or alternative pathways: the retention of residual leaf area after defoliation, a high density of residual buds, bud activity, and the allocation of assimilate to new leaves rather

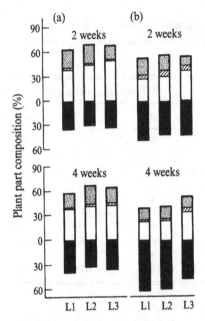

Figure 4.2. Relative composition of plant parts (root ■, stubble □, stem ▨, and leaf ▦) of (a) *P. malacophyllum* and (b) *P. wettsteinii* under cutting frequencies of 2 or 4 weeks and grown in light levels of 20 (L1), 50 (L2), and 100% (L3) full sunlight. (From Wong, 1993.)

than to roots and crown. Richards (1984) contrasted the superior resistance of *Agropyron desertorum* to defoliation, relative to that of the more vulnerable *A. spicatum*. He associated this differential response with the continued allocation of carbon to the roots of the latter following clipping, whilst the former invested carbon in shoot regrowth which restored photosynthetic capacity more rapidly. Mortality of *A. spicatum* roots was then greater in the subsequent winter than that of *A. desertorum*.

Similarly, Wong (1993) contrasted the defoliation responses of an erect grass, *Paspalum malacophyllum,* with that of a prostrate grass, *P. wettsteinii.* The latter unexpectedly proved less resistant to frequent defoliation than the former, in terms of persistence and vigour of regrowth, especially under shaded conditions, as would occur if these plants were grown as an understorey in plantation agriculture. The relative allocation of carbon to roots under 2-week and 4-week cutting systems (Figure 4.2) was greater in the sod-forming *P. wettsteinii* than in the bunch grass *P. malacophyllum.* However, this type of comparison fails for the grazing resistant *Cenchrus ciliaris* and the vulnerable *Themeda triandra* (Hodgkinson *et al.*, 1989).

Duration of carbohydrate starvation

The crucial question is the duration of the period after severe defoliation before shoots become photosynthetically self-supporting. This period is zero for swards with active meristems and green leaves below the level of defoliation, and extends from zero to 3 to 6 days if defoliation is extreme. The potential loss of growth in this period needs to be viewed in the context of the reduced herbage utilization and the decreased nutritive value associated with systems of lenient defoliation directed to reserve accumulation.

There is a great body of scientific evidence which indicates an imperfect or minor relationship between TNC and the overall rate of regrowth after defoliation (for example: Adegbola, 1966; Carlson, 1966; Humphreys & Robinson, 1966; Jones & Carabaly, 1981; Richards & Caldwell, 1985; Volenec, 1986; Barnes, 1989; Danckwerts & Gordon, 1989; Hodgkinson *et al.*, 1989). There is also good evidence from ^{14}C studies that the rate of regrowth is enhanced by the reassembly of TNC in new shoots in the immediate, short phase after severe defoliation (Steinke & Booysen, 1968; Beaty *et al.*, 1974). It is of interest that Davidson & Milthorpe (1966a) found a positive relationship between the rate of leaf expansion and the total soluble carbohydrate content of the leaf bases when TNC was varied by placing the plants in the dark but not when it was varied by defoliation and subsequent growth. An analogous situation may arise when herbage plants are grown in the shaded conditions of silvipastoral systems. A positive relationship is shown (Figure 4.3) between the etiolated regrowth of *P. malacophyllum* and *P. wettsteinii* and the regrowth of shoots to day 3 and to day 7 after defoliation when plants had received 20 or 50% full sunlight in south east Queensland (Wong, 1993). This study raises the question of whether TNC may be a more significant factor in the management of pastures grown in shade, where carbohydrate is in more limited supply, than in the management of pastures grown in favourable radiation regimens.

4.3 Frequency and height of cutting

The second major emphasis of early studies of pasture response to defoliation related to the effects of variation in the frequency and height of cutting. This led to a voluminous literature of local and cultivar responses to cutting management. Early studies at Aberystwyth (Fagan & Jones, 1924; Stapledon & Davies, 1930) emphasized how leafiness varied according to cutting management and cultivar. A general finding was that infrequent cutting maximized dry-matter yield, especially of bunch grasses, but

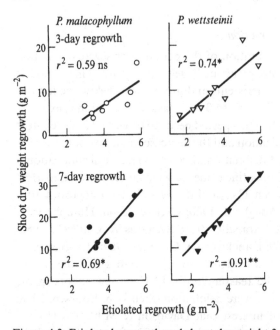

Figure 4.3. Etiolated regrowth and shoot dry weight 3 and 7 days after defoliation of *P. malacophyllum* and *P. wettsteinii* grown in 20 and 50% full sunlight. (From Wong, 1993.)

decreased leafiness, N concentration, and nutritive value, and increased structural components; a compromise might be sought which maximized yield of digestible nutrients (Davies & Sim, 1930; Cashmore, 1934; Bredon & Horrell, 1963). Low cutting height increased the percentage utilization of nutrients but decreased persistence, depending on plant habit (Albertson, Riegel & Launchbauch, 1953; Branson, 1956), whilst frequent cutting might extend the seasonal growth period relative to infrequent cutting. Competitive relations between components of mixtures were varied by cutting management (Datzenko & Ahlgren, 1951; Gervais, 1960) and the relative resistance of cultivars to severe defoliation, in terms of persistence and growth, was characterized. These findings did not lead to further conceptual development or to wider field application for two reasons.

Harvested yield

Harvested yield was measured, but usually the amount of material remaining after each cut and its morphological characteristics were not recorded. The data were not therefore susceptible to growth analysis and the seasonal record of harvest for different treatments was confounded in an unidentifiable way with differences in residual stubble and did not necessarily reflect accurately treatment effects on growth.

Grazing and cutting

The results from these early cutting trials were applicable to 'cut and remove' feeding systems which are often used for convenience in mixed farming systems in the tropics and to similar systems occasionally used elsewhere in mechanized 'green-lot' systems, especially for dairying. However, the results had a limited pertinence to pastures managed for hay or silage and to the bulk of the world's pastures where ruminants feed by grazing and browsing, which have effects distinct from that of cutting. Apart from the obvious differences in the type and extent of trampling (Edmond, 1966) and the return of excreta (Doak, 1952; Jones & Ratcliff, 1983), grazing is selective spatially, either in the macro-dimension of the concentration of defoliation in patches (Watkin & Clements, 1978; Mott, 1985) and on whole plants of different species, or in the micro-dimension of individual tillers or shoots and the individual organs comprising these. The response of plants to non-selective cuts at a specified height is different from that of grazing, and may result in more seasonally restricted growth (Orr *et al.*, 1988).

4.4 Leaf area index and growth analysis

The next phase of the thinking about defoliation responses emerged from the development of growth analysis. V.H. Blackman conceived plant growth as analagous to the growth of compound interest; relative growth rate related to previous plant size and F.G. Gregory (1938) and others differentiated it as the product of the efficiency and the size of the photosynthetic system, as expressed in the ratio of leaf area to plant weight. D.J. Watson's seminal contribution was the invention of the leaf area index (LAI, the area of leaf supported by unit ground surface), which expressed growth rate and crop yield in absolute terms when multiplied by net assimilation rate. The distribution of assimilate to the organs of interest then determined cereal grain or tuber yield (Watson, 1947, 1952, 1968); leaf was the predominant interest for pastures.

R.W. Brougham (1956, 1958) applied these concepts to the regrowth of pastures after cutting; height of cut determined the duration of the phase until the leaf surface was sufficiently restored to intercept radiation. This led to the idea that pastures might be utilized to avoid radiation wastefully

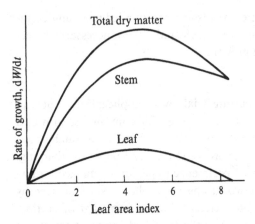

Figure 4.4. The general relationship between the rate of increment (dW/dt, where W is shoot mass and t is time) of weight of leaf, stem and total shoot DM and the LAI of *T. subterraneum*. (From Davidson & Donald, 1958.)

striking the ground surface when this energy might otherwise be captured to augment the photosynthetic output. Critical LAI was suggested as the value at which 95% incident radiation was intercepted. J.L. Davidson and C.M. Donald (1958), working with *T. subterraneum*, demonstrated an optimum LAI (value 4–5) which maximized pasture growth (Figure 4.4); pasture growth decreased by *c.* 30% as LAI increased to 8.7 and incorporated senescing leaves. They suggested that the optimum LAI increased as radiation conditions became more favourable, since fewer leaves would be below the compensation point than at lower levels of illuminance, and that the effect of defoliation depended upon the LAI at which defoliation occurred and the value to which it was reduced.

Saeki (1960) recorded the passage of light through the sward, and calculated the extinction coefficient, which expressed the capability of a unit leaf surface to intercept radiation. Warren-Wilson (1961) related optimum utilization of incident light to LAI, leaf inclination, orientation, height distribution and dispersion; this depended upon most of the foliage receiving uniform light of low intensity with the lowermost layer at compensation point. An optimum LAI might be indicated (Murtagh & Gross, 1966), or growth rate might tend to an asymptote at high values of LAI (Brougham, 1956; Brown & Blaser, 1968) if respiration was proportional to the rate of photosynthesis (Ludlow & Charles-Edwards, 1980). LAI is best expressed as the sum of all green surfaces, including green stem.

There were three main flaws in the applicability of these studies to the field.

Sward growth and utilized yield.

Some crop physiologists sought to manipulate LAI in order to maximize pasture growth, but the first objective for grazed pasture is to optimize the yield of herbage nutrients ingested. Further, this does not derive directly from the harvested DM yield, since management directed to maximizing DM yield leads to the production of herbage with a high content of stem, senescing leaf and lowered nutritive value (Parsons *et al.*, 1983a, b).

Senescence

Increase in sward DM represents the net balance between new growth and senescence; early studies gave little attention to the latter process. Senescence is visible when drought or frost cause a cessation of growth or if pastures are grossly undergrazed; it was not universally appreciated that in fact the highest rates of senescence occur when pastures are actively growing (Humphreys, 1966b; Wilson & Mannetje, 1978; Grant *et al.*, 1983; Leafe & Parsons, 1983). Uneaten herbage exports some nutrients and energy to other organs but the major part enters the pool of litter. Defoliation practice should be therefore directed to synchrony of growth and consumption.

Other constraints

Management targeted towards maximizing light interception by the sward is irrelevant in circumstances where environmental factors other than light control growth of the sward. The output from irrigated pastures, pastures of the humid tropics and from pastures in humid temperate pastures during the warmer months may be enhanced by appropriate manipulation of the leaf surface which focuses on radiation use. The growth of other types of pasture is often constrained by shortages of moisture and of nutrients which limit the extent of the green surfaces which can be maintained. Swards with high LAI wilt earlier than heavily used swards; swards with low residual LAI after grazing exhibit increased leaf expansion and net assimilation rate and a homeostasis tends to equilibrate shoot yield over a wide range of management conditions (Humphreys, 1966a).

4.5 Characterization of defoliation

In this chapter defoliation has been used in the generic, applied sense of shoot removal rather than in the exact sense of detachment of lamina. The amount, age and type of tissue removed and the timing of removal are the significant variants. The degree of removal determines the instantaneous situation with respect to residual LAI, which influences whether shoots continue to be photosynthetically self-sufficient and whether defoliation induces an acute shortage of carbohydrate. The removal of young leaves, which have a higher photosynthetic capacity than aged leaves, is detrimental to sward growth (Jewiss & Woledge, 1967; Whiteman, 1970), and grazing animals selectively remove accessible young leaf. Removal of the upper canopy may expose older leaves in the lower canopy; if these have developed in shade they may be expected to show high specific leaf area but reduced photosynthetic capacity (Ludlow & Wilson, 1971b). Retention of the expanding leaf bases of grasses (Davidson & Milthorpe, 1966a) or the unfolded lamina of legumes (Culvenor *et al.*, 1989a) assists rapid restoration of the leaf surface.

The removal of an elevated shoot apex may be detrimental to sward recovery, especially if there is a shortage of residual basal shoots or active basal buds. F.A. Branson (1953) directed attention to the propensity of some grasses (*Panicum virgatum, Agropyron smithii*) to elevate vegetative apices which were still differentiating leaves into the grazing zone, whilst other grasses (*Poa pratensis, Buchloe dactyloides*) maintained apices at or below the soil surface, and were resistant to grazing. This type of finding formed a suggested basis for grazing management (Scott, 1956; Booysen, Tainton & Scott, 1963), but is also applicable to species evaluation (Muldoon & Pearson, 1979b; Clements, 1989). The significance of meristem removal is expanded later in the chapter. Defoliation of reproductive shoots impairs seed production but may enhance forage quality (Hoogendorn, Holmes & Chu, 1992).

4.6 Current emphases

Current perceptions of the responses of forage plants to defoliation and of the plant characteristics which assist the rapid replacement of the tissues removed have been well reviewed by Richards (1993) and Parsons (1993). These build onto concepts discussed so far, especially with respect to (i) the carbon balance; (ii) the dynamics of the leaf surface and the associated activity of the meristems; and (iii) biological N fixation.

4.6.1 The carbon balance of swards

The dichotomy between growth and utilization, the need to take account of herbage senescence, and the factors influencing carbon allocation to different plant organs were mentioned earlier.

The emphasis on attaining high levels of utilization derives in part from the recognition that high levels of herbage accumulation, associated with maximizing herbage growth, result in low tiller density, depressed nutritive value, and wasteful herbage senescence (Stakelum & Dillon, 1989). Thus, in Germany Koblet (1979) recorded high efficiency of growth on dense, intensively grazed swards containing a predominance of young leaf material; net herbage growth rates were close to maximal at LAI 2–3.

Parsons & Penning (1988) at Hurley, UK proposed a simple model which illustrates the components of herbage accumulation in rotationally grazed swards of *L. perenne* cv. S23 subjected to varying durations of regrowth after severe grazing. From other data gross photosynthesis (Figure 4.5a) was found to increase rapidly as the leaf surface was restored, the slope diminishing with time, whilst shoot tissue production reached an asymptote as respiration and root growth accounted for a large proportion of gross photosynthesis. Senescence lagged behind net photosynthesis, so that the instantaneous growth rate (dW/dt, where W is shoot mass and t is time) which records the difference between these components, appears as an early bulge in Figure 4.5a and an early peak in Figure 4.5b. Thus, there was a period in relatively early regrowth when high rates of tissue production were associated with a low rate of tissue loss, and which contributed to the development of the sigmoid function for accumulation of shoot mass (W). However, it is best to consider the average growth rate, which is defined as the net increase in the weight of the sward ($W-W_0$) divided by the time that has elapsed since grazing, since it is desirable to harvest the pasture when the average growth rate is maximal. At an early stage ((1) in Figure 4.5b) both average and instantaneous growth rate were increasing, but in later stages ((2) and (3) in Figure 4.5b) variation in average growth rate was damped by earlier high values of dW/dt, the maximum value for average growth rate occurring before the attainment of ceiling yield.

These results may be compared with other studies at Hurley involving infrequent cutting or continuous grazing directed to maintaining an LAI of 1.0 or 3.0 (Table 4.2). Gross photosynthesis was least under heavy continuous grazing; in Table 4.2 shoot and root respiration were simplistically set at 50% P_{gross}. The actual yield of swards infrequently and severely cut was greater than that of swards with a short recovery period after grazing (12–13 days), but similar to those with a medium (19–23 days) and long (30–34 days) recovery period, when yield of these grazed swards was estimated by cutting the recovery growth. However, the actual harvesting of herbage by grazing sheep was impaired in the medium and long recovery treatments, associated with the stemminess of the pasture, so that the

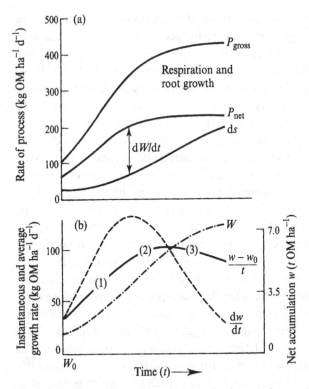

Figure 4.5. Effect of duration of regrowth after grazing on (a) rate of gross photosynthesis (P_{gross}), tissue production (P_{net}) and death (dS) and on (b) instantaneous growth rate (dW/dt), weight of the crop (W) and the average growth rate ($W-W_0/t$). See text for (1), (2) and (3). (From Parsons & Penning, 1988.)

utilized yield of all three rotationally grazed treatments was similar to that of the severe continuous grazing treatment, despite the decreased shoot tissue production of the latter, which displayed the highest percentage efficiency of harvest. Further modelling (Parsons, Johnson & Harvey, 1988) suggests that the characterization of production of swards under both continuous and intermittent defoliation and the rationalization of their management are best done on the basis of the average LAI attained; this generalization is supported in south east Queensland by Ludlow & Charles-Edwards (1980). The use of residual LAI to characterize systems of rotational grazing systems is insecure, since the outcome also depends upon the recovery interval. The objective is to find for different sward types, localities and seasons the optimum balance between photosynthesis, gross tissue production, herbage intake, and plant senescence.

Table 4.2 L. perenne *production (t OM ha⁻¹ 180 d⁻¹) under cutting, continuous grazing (LAI 1.0 (H) and 3.0 (L)), and rotational grazing for various intervals of recovery.*

Component	Infrequent severe cutting	Continuous grazing		Rotational grazing interval		
		H	L	Short	Medium	Long
(a) Gross photosynthesis	60	38	54	49	62	69
(b) Shoot tissue production	30	19	27	24	31	35
(c) Yield						
Cut	12–15	–	–	10	13	16
Grazed	–	9.6	6.9	9.6	9.4	10.2
Efficiency of harvest (%)	40–50	51	26	40	30	30

Source: Parsons & Penning (1988).

One noteworthy phenomenon is 'compensatory' photosynthesis which has been reported for several species following defoliation; rates of photosynthesis are higher than those of leaves of the same age on undefoliated plants (Richards, 1993), and this also extends to the relative photosynthesis of new leaves produced after defoliation. Enhanced photosynthesis usually arises from the increased photosynthetic capacity of the mesophyll, but may also be associated with increased stomatal conductance (Gifford & Marshall, 1973). This appears to be due less to a direct change in source–sink relations, but rather to substances (such as cytokinins) transported from the roots to the leaves of defoliated plants (Carmi & Koller, 1979). It results in either a delay in leaf senescence or a rejuvenation of leaf tissue (Hodgkinson, 1974).

Compensatory photosynthesis aids the carbon balance of defoliated swards; a more significant factor was mentioned earlier: the preferential allocation of carbohydrate to leaf meristems (Richards, 1984). The notion that preferential transport of assimilate to roots (rather than shoots) confers intolerance of defoliation represents a changed mind-set; paradoxically, it is the more rapid recovery of carbon self-sufficiency consequent upon allocation to leaves which leads to the eventual restoration of the root system. The flow of carbohydrate may be interpreted in terms of source–sink relations and the relative strength of sinks (Bucher, Mächler & Nösberger, 1987). The allocation of carbon is partially modified by the phytochrome system and defoliation which increases the red:far-red (R:FR) ratio of light striking the base of the sward may be expected to increase movement of assimilate to leaves and increase branching and to reduce transport to roots and nodules, as in *T. repens* (Robin *et al.*, 1993).

The degree of interdependence of tillers or shoots of differing rank influences the ecological success of grazed plants. Thus, for the perennial grasses *Paspalum plicatulum* and *Shizachyrium scoparium* in Texas the carbon import of the daughter tillers from the main tiller increased if defoliated or shaded; this phenomenon occurred within 30 min. of treatment, indicating that a functional transport system was in place (Welker *et al.*, 1985). For *T. repens* Chapman, Robson & Snaydon (1992) observed that carbon moved freely in both directions between the parent stolon and the branches; the branches normally exported more C to the parent plant than they received. Old branches were relatively self-sufficient for carbon, and were unaffected by the defoliation of the parent stolon, whereas the growth of young branches was reduced. Defoliation of branches was buffered by transfers from the parent stolon, and it is suggested that a highly branched plant may exhibit relatively stable growth when subjected to the spatially heterogeneous defoliation which occurs under grazing.

4.6.2 Dynamics of the leaf surface and meristematic activity

Restoration of the leaf surface after defoliation and its subsequent maintenance depend upon the balance within the development sequence: residual density of buds and their activation to determine shoot density, the rate of leaf appearance on individual shoots, the elevation of leaves in the canopy and the size which the lamina attains, the photosynthetic capacity of the leaves during their life, and the rate of their senescence.

Many scientists have focused on the activation of buds as the key to the rate of recovery growth. Leach (1970) related the regrowth of *M. sativa* to the density of basal buds and the time when each elongated, and discounted residual LAI and TNC as having limited effects; Busso, Richards & Chatterton (1990) suggested high TNC of range grasses facilitates regrowth only when meristematic activity is high. Defoliation releases lateral buds from their suppression by established shoots, and the concept of apical dominance (Phillips, 1975; Knox & Wareing, 1984), associated with the balance and supply of specific hormones, is invoked to explain in part the accelerated rate of shoot appearance which follows defoliation (Muldoon & Pearson, 1979a). In *P. maximum* var. *trichoglume* young (but not old) flowering shoots are inhibitory (Humphreys, 1966b), whilst Jewiss (1972) and Clifford (1977) emphasize the negative effect of stem extension linked to inhibitory levels of auxin in elongating stem internodes. An optimum concentration of cytokinin activates lateral buds (Nojima, Oizumi & Takasaki, 1985) whose subsequent growth is promoted by gibberellin (GA_3) (Oizumi *et al.*, 1985). Light quality, as indicated by the red:far-red ratio, is increased by defoliation, since the far-red wavelengths which limit bud activity (Deregibus *et al.*, 1985) at the base of tall swards (Barthram, Grant & Elston, 1992; Chapman & Lemaire, 1993) are less evident.

The beneficial effects of high rates of tiller appearance need to be qualified by reference to the disadvantages of synchrony; the absence of daughter tillers when the initial flush of tillers of *Themeda triandra* was removed promoted its demise, under conditions where the asynchronous tillering of *Cenchrus ciliaris* and *Heteropogon contortus* contributed to the survival and growth of these species (Mott *et al.*, 1992).

High producing grazed swards of temperate grasses have a characteristic live leaf number per shoot, and a leaf appearance interval which is strongly correlated with temperature. These values are, for example, 2.5 green leaves per tiller and 220 degree days for *Festuca arundinacea* and 3.0 green leaves and 110 degree days for *L. perenne* (Chapman & Lemaire, 1993). Defoliation interval directed to minimizing leaf senescence may be modified by this type of information. Compensatory effects are evident in the components of the leaf surface; heavy grazing increases the tiller density of *L. perenne* (Curll & Wilkins, 1982; Binnie & Chestnutt, 1994) but also leads to reduced leaf length. These types of analysis may be used to clarify the competitive relations of the components of mixtures. Parsons, Harvey & Woledge (1991) noted an increase in *T. repens* (clover) and a decrease in *L. perenne* (grass) during summer grazing. This was associated with a greater number of foliated nodes per stolon (or tiller) in clover than in grass, and a lesser investment of DM in clover lamina was compensated by greater specific leaf area in clover. Grass leaves suffered their first grazing earlier than clover leaves, and since clover leaves escaped grazing by appearing and expanding close to the soil surface below sward height, a greater proportion of clover leaf area was in the more productive growing-leaf category and was only removed at a later stage by grazing when petioles extended the laminae near the top of the canopy. These types of responses were evident over a range of defoliation intensities and emphasize the significance of the response of the leaf surface to grazing in determining ecological success.

4.6.3 Biological N fixation

Severe defoliation reduces legume N fixation (Butler, Greenwood & Soper, 1959); its effects are evident in premature nodule senescence, reduced

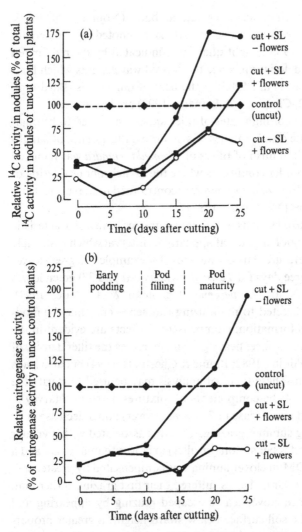

Figure 4.6. Effects of cutting, stubble leaf (SL) removal (–) or retention (+) and flower removal or retention in *Macroptilium lathyroides* on (a) supply of ^{14}C-labelled assimilate to nodules and (b) nitrogenase activity. (From Othman *et al.*, 1988.)

nodule size and delayed nodule initiation (Whiteman, 1970). Biological N fixation is closely related to the rate of legume growth, as discussed in chapter 3, and there is strong associative evidence that in effectively nodu-lated plants this is essentially mediated via the supply of assimilate to nodules. This is well illustrated (Figure 4.6) by a study (Othman, Asher & Wilson, 1988) in which *Macroptilium lathyroides* was cut at 14 cm (8th node) when plants were at an early podding stage, and the stubble leaves were removed or not. The ^{14}C activity in the nodules was reduced to 25–40% of the uncut control 2–3 hr after cutting, and fell further in the most severely defoliated treatment; ^{14}C levels were restored after 25 d if stubble leaves were retained (Figure 4.6a). Nitrogenase activity in the nodules (Figure 4.6b) showed a close parallelism with the supply of assimilate. The competitive sink of flowers and developing pods greatly reduced N fixation, and also appeared to compete for N with axillary buds. There are other studies (Kouchi & Nakaji, 1985, for soybean) which show a close association between current photosynthesis and nodule metabolism.

A notable discovery (Hartwig, Boller & Nösberger, 1987) was that nodule diffusion resistance to oxygen increased markedly after defoliation, and this anoxia limited nitrogenase-linked respiration, as demonstrated in Table 4.1. Hartwig *et al.* (1990) then went further to claim that 'lack of photosynthates is not the immediate cause of the decline of nitrogen-fixing activity after defoliation', since *T. repens* nodule activity did not increase if roots were bathed in glucose; however, TNC levels were higher than in many legume situations. It does appear that defoliation leads to a closure of the open intercellular spaces between the thick-walled cells of the cortex; this arises from an increase in the amount of glycoprotein, which draws water into the cells, and the expansion of the cells within the inner cortex may restrict the diffusion of oxygen by a factor of four (Minchin, 1994). The mechanisms involved in the response of N fixation to defoliation are complex, but there is evidence from many studies that management which impairs the flow of assimilate to the nodule sink impairs N fixation.

A final comment is that if the optimization of biological N fixation is the principal *raison d'être* for growing legumes then a system of grazing management which maximizes legume growth during the growing season has much to commend it. This occurs naturally in legume–grass mixtures where the legume tends to be rejected by grazing animals during the wet season (Böhnert, Lascano & Weniger, 1985) and is selectively eaten during the dry season, when the companion grass has greatly reduced nutritive value.

4.7 Conclusion

This developmental review indicates the considerable distance travelled by grassland scientists in arriving at the current emphases described above. The predictability of sward response to defoliation has increased substantially for several pasture types and environments, but the number of processes involved and the effects of abiotic influences make extrapolation to other situations hazardous. Perhaps the greatest weakness, and one which has received little attention in this review, is the prediction of plant mortality (Mott, McKeon & Day, 1993); it is known that plants die before TNC (Wong, 1993) or other resources are exhausted, but the mechanisms are unknown.

The influence of sward surface height on production is discussed further in chapter 6. Plant evaluation and management decisions are benefiting from a focus on the rapid refoliation consequent upon the preferential allocation of C and N to shoot meristems, and the maintenance of active meristems and leaf area below the height of defoliation (Parsons, 1993).

The Grassland Research Institute at Hurley was closed by the UK Government in 1990. One of the enduring monuments to its influence on grassland improvement is the series of studies undertaken there which modelled in new and elegant ways sward production, partitioning of assimilate senescence, herbage removal and animal production into a comprehensive whole (Orr *et al.*, 1988, 1990, 1995; Parsons, Johnson & Harvey, 1988; Parsons, Johnson & Williams, 1988; Parsons & Penning, 1988; Penning *et al.*, 1991). These studies also provided a physiological basis for understanding the predator/prey interactions of foraging theory which lead to changes in the composition of mixed swards and of animal diet (Parsons, Harvey & Johnson, 1991; Parsons, Harvey & Woledge, 1991). Critical studies at this level are unavailable for tropical pastures.

5
Grassland ecology

5.1 Grassland climax and succession

5.1.1 Introduction

Grassland ecology attracts scientists interested in the interplay of the biotic, climatic and edaphic components of the environment as these modify the dynamics of grassland communities. Understanding of the processes involved provides conceptual bases for the management of natural plant communities and may underpin the efforts of those concerned with the use, improvement and conservation of 'rangeland' resources. These ecological insights are also directed to manipulating the botanical composition of planted pastures, especially as directed to the maintenance of a herbage legume component or to the control of weeds.

This topic is wide and this chapter is therefore focused on the main areas of controversy: the demise of Clementsian succession, the rise of the state-and-transition model, the replacement of the evangelistic piety of range management by more rational and quantitative assessments of land condition, and the advances in describing the processes of plant adaptation and population change.

5.1.2 The Clementsian monoclimax and its modification

The most influential twentieth century concept in grassland ecology was developed by F.E. Clements and his predecessors (Clements, 1916); they viewed vegetation as a dynamic (and not static) entity in which 'succession is the universal process of formation development . . . each climax formation is able to reproduce itself, repeating with essential fidelity the stages of its development'. These stages or seral units consisted of plant communities which could be used as plant indicators (Clements, 1920) of

129

anthropogenic interventions such as variation in the intensity of grazing and which developed along an inexorable (but still reversible) linear path to a monoclimax which reflected the climate of the region. Clements (1920) drew on the earlier work of J.G. Smith in south-western USA to indicate the regression from *Andropogon* dominance to *Aristida* to *Hilaria* and *Bulbilis* as grazing pressure increased. A.W. Sampson (1919) further developed the relations of plant succession to range management. Bare ground was colonized by ruderals, which gave place to seral grassland stages as organic matter accumulated, and these were eventually replaced by taller bunch grasses.

Clements' influence at the University of Nebraska persisted and collaborative work with other institutions such as Kansas State College and North Dakota Agricultural College made the US Great Plains region (Joyce, 1993) the dominant centre for the propagation of these theories, which were taken up next in Canada (Coupland, 1952), Australia and southern Africa, where J.F.V. Phillips influenced generations of students. All plant communities are both climatically and edaphically determined; Clements recognized edaphic subclimaxes and different populations within an area could be regarded as sub-units of the same climax. Whittaker (1953) enumerates the proliferation of some 36 terms with '-climax' as a suffix which attempted to accommodate the diverse viewpoints of ecologists. Of these the disturbance climax or disclimax and its variant, the fire climax, are of greatest utility in grassland ecology. The application of the concepts of succession and climax to range management developed as the status of plant indicators (Sampson, 1939), the definition of site production potential (Humphrey, 1949) and erosion condition were elaborated; Dyksterhuis's (1949, 1952) explication of land condition in the quantitative terms of present species composition as a proportion of species composition of the climax was then widely applied. A coherent body of teaching became enshrined in texts on range management such as Weaver & Clements (1938), Stoddart & Smith (1943), Sampson (1952), Humphrey (1962) and Heady (1975).

5.1.3 Grassland succession

The linear succession model (Figure 5.1) arrays all vegetation states in a single continuum in which the early successional states, which reflect excessive grazing and poor land condition, may be improved through fair, good and excellent condition states as the climax is approached through conservative grazing management (Westoby, Walker & Noy-Meir, 1989). This

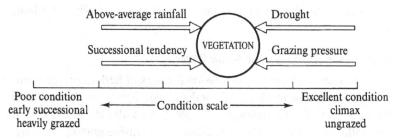

Figure 5.1. General scheme of the range succession model, as modified by rainfall variability. (From Westoby *et al.*, 1989.)

model has been modified to indicate that seasonal rainfall variation operates in a similar way; in above-average rainfall years grazing pressure is usually reduced and forage allowance increased if animal density is relatively constant, and the reverse applies in drought years.

Two illustrations of successional trends are given.

The Transvaal highveld

In the Transvaal highveld of South Africa the following stages operate after disturbance (Davidson, 1962):

(i) Ruderal stage: mainly dicotyledonous annual weeds such as *Acanthospermum brasilum* and *Tagetes minuta*, but including grasses such as *Eleusine indica, Panicum laevifolium, Setaria pallide-fusca* and the sedge *Cyperus esculentus*.

(ii) First grass stage: stoloniferous grasses, principally *Cynodon dactylon*, sometimes with *Paspalum commersonii* and the naturalized species *P. dilatatum, P. notatum* and *Pennisetum clandestinum*.

(iii) Second grass stage: bunch grasses, mainly *Eragrostis curvula, E. racemosa, E. plana, E. gummiflua, Sporobolus pyramidalis* and *Rhychelytrum repens*; the woody subshrub *Stoebe vulgaris* occurs occasionally.

(iv) Third grass stage: The tall bunch grass *Hyparrhenia hirta* dominant with occasional dicotyledenous species.

(v) Grassland subclimax: tall bunch grasses such as *Trachypogon spicatus, Tristachya hispida, Elyonurus argenteus, Digitaria tricholaenoides, Heteropogon contortus* and *Themeda triandra*.

(vi) Scrub climax: various scrub first species, arrested by fire and grazing.

Increasing SR from 1.2 b ha^{-1} to 2.3 b ha^{-1} caused a regression in species composition from the grassland subclimax to the second and first grassland stages, which were also more resistant to artificial trampling (Gillard, 1969).

The southern tablelands of New South Wales

The second example shows the successional stages of herbaceous communities in *Eucalyptus melliodora – E. blakleyi* woodlands on the southern tablelands of New South Wales (lat. 35–36° S, rainfall 580 mm, *c.* 600 m altitude) following clearing, the imposition of increasing grazing pressure, and the application of superphosphate (Moore, 1967). This region has an erratic rainfall which on average is distributed evenly between summer and winter.

The grassland climax species (Stage I, Figure 5.2) are tall (0.9–1.2 m) perennial tussock grasses ('decreasers') which grow in the warm season, partly on residual soil moisture accumulated from winter rains. The introduction of domestic grazing animals leads first to the loss of *Themeda triandra* and the appearance of *Stipa falcata* (Stage II); heavier grazing (2.5–4 sheep ha^{-1}) results in the increase of short (5–15 cm) perennial grasses ('increasers') which grow in the cool season (Stage III). The latter become dominant, but the opening of the sward under heavy grazing leads to the invasion of dwarf native species from more arid communities which can utilize rainfall in the warm season (Stage IV). This is followed rapidly (Stage V) or coincidentally by the incursion of exotic Mediterranean annual plants ('invaders'), including the legume *Trifolium glomeratum*. The application of superphosphate and the augmented carrying capacity (5–10 sheep ha^{-1}) leads to the loss of perennial grasses and thus complete replacement (Stage VI) by nitrophilous cool season annuals such as *Vulpia bromoides* and *Bromus rigidus* allied with warm season grazing resistant grasses such as *Chloris truncata* and *Eragrostis cilianensis*. This example can also be fitted to the 'state-and-transition' model which is discussed subsequently.

5.1.4 Succession and soil fertility

Successional changes are linked to soil fertility. Figure 5.2 reflects succession on a relatively infertile podzol (spodosol); in the same region succession on less heavily leached soils occurs similarly without the addition of superphosphate (Moore, 1967). The climax grass *T. triandra* gives low sheep production since its protein content is low in winter and early spring, and it is difficult to estimate carrying capacity since the community is so unstable near the climax under grazing. The maximum soil nitrate-N (0–10 cm) at the end of summer was 1 ppm for Stage I; this increased to 2, 4 and

CLEARING

I   *Themeda triandra – Stipa aristiglumis – Poa caespitosa*
(Tall warm season perennial tussock grasses)

GRAZING

II   *Stipa aristiglumis – Poa caespitosa – Stipa falcata*

III *Stipa falcata – Danthonia carphoides – Danthonia auriculata*
(Short perennial cool season grasses)

IV   *Danthonia carphoides – Danthonia auriculata*
(Dwarf cool season perennials)

Native species { *Enneapogon nigricans, Chloris truncata,*
from more arid { *Tripogon loliiformis, Panicum effusum,*
communities { *Vittadinia triloba, Euphorbia drummondii,*
{ *Convolvulus erubescens*
(Dwarf warm season species)

GRAZING

V   *Danthonia carphoides – Danthonia auriculata*

Mediterranean { *Trifolium glomeratum, Vulpia bromoides,*
annuals { *Bromus* spp., *Hordeum leporinum,*
{ Composites, *Erodium cicutarium,*
{ *Aira caryophyllea*
(Exotic cool season annuals)

Enneapogon nigricans, Chloris truncata, Panicum
effusum, Vittadinia triloba, Eragrostis brownii
(Short warm season perennials)

GRAZING AND SUPERPHOSPHATE

VI *Vulpia bromoides, Bromus rigidus, Erodium cicutarium,*
Trifolium glomeratum, Cirsium vulgare
(Exotic cool season annuals)

Chloris truncata, Eragrostis cilianensis,
Chenopodium carinatum, Polygonum aviculare
(Warm season species)

Figure 5.2. Species changes in the herbaceous communities of woodlands on the southern tablelands of New South Wales in response to clearing, increasing grazing pressure, and superphosphate application. (From Moore, 1967.)

36 ppm respectively for Stages II, III and VI. Moore (1967) regards the cool season communities as nitrogen exploiting systems and the end-point of further grazing and superphosphate application is the ingress of unpalatable nitrophilous plants such as *Silybum marianum, Onopordum acanthium, Carduus pycnocephalus* and *Hordeum leporinum* and soil nitrate-N in excess of 100 ppm.

The reverse situation of colonization of bare ground and the interaction of subsequent succession with soil fertility is illustrated from a study (Davidson, 1962) of Transvaal highveld at Frankenwald (lat. 26° S, 760 mm annual rainfall, 1500 mm altitude). Land which had been cropped to maize was retired from cultivation, and fertilizer was applied. The successional stages of this grassland were described at the beginning of section 5.1.3. Ruderals dominated for the first two years, and Table 5.1 shows the yield and botanical composition eight years after cessation of cultivation. Superphosphate alone (which contained calcium sulphate as well as calcium monophosphate) had little effect on successional development. The notable effects evident are the arresting of the third grass *Hyparrhenia hirta* stage in the two plus N treatments, the persistent first grass stage *Cynodon dactylon* in the PN_5 treatment and the higher yield of the latter treatment.

Many grasses of the southern African flora do not flourish under conditions of high N supply, and Grunow, Pienaar & Breytenbach (1970) have categorized the main grasses as decreasers or increasers according to N availability. This response has been attributed to intolerance of the ammonium ion, but it may also be related to relative competitive capacity, as discussed later. The lower successional stages are marked by a lesser accumulation of litter and a more rapid cycling of N under the grazing regimes imposed.

The obverse relationship of succession to soil fertility applies where lower successional stages are associated with soil erosion and the loss of topsoil in which nutrients are concentrated.

5.2 State-and-transition models

5.2.1 Inadequacies of climax and linear successional theories

Clementsian succession worked well in guiding range management in the Great Plains region of America where the concept developed; its inadequacies have become more evident when applied in wider contexts. These are summarized:

Table 5.1 *Previous fertilizer treatment, percentage composition, and yield of different seral stages of Transvaal highveld*

Seral stage	Fertilizer treatment			
	Control	P	PN_1	PN_5
Ruderals Dicotyledonous species	8	5	8	2
First grass stage Cynodon dactylon	17	9	8	30
Second grass stage Eragrostis spp.	51	50	71	57
Third grass stage Hyparrhenia hirta	24	37	13	11
Total herbage yield (kg ha^{-1})	2360	1680	2660	5250

Control, no fertilizer, P, 224 kg ha^{-1} yr^{-1} single superphosphate; PN, as for P plus 22 kg N ha^{-1} yr^{-1}, PN_5 as for P plus 110 kg N ha^{-1} yr^{-1}.

The status of pristine vegetation

In areas long used for intensive grazing it is difficult to establish the nature of the climax, and the comparison of existing vegetation with this notional community may have little relevance to correct management. Relict areas, e.g. cemeteries and natural wildlife parks, have been used to establish the composition of the climax (Parker, 1952) and the establishment of stock exclosures has been advocated as a means to establishing ecological benchmarks. These fail if the grassland inside the exclosure is clearly in a moribund, unproductive condition, as occurs for example in the *Astrebla* grassland of western Queensland where reduced grazing accelerates plant senescence (Orr, 1980).

Reversibility of regression

There are many circumstances where Clements' reference to 'the irresistible impulse towards the climax' is misplaced. The distance away from the climax is usually predictably increased in a non-linear manner with increasing grazing pressure; the converse does not necessarily occur or demonstrates hysteresis by a different pathway. In the north Pennines of the UK the removal of stock from *Nardus* hill pastures dominated by unpalatable species does not give reversion to palatable species (Rawes, 1981). The perennial grasses in the Mediterranean climate of California which were lost to annual grassland by grazing are not restored by stock exclusion (Heady, 1958).

Climax productivity.

The climax grasses of the Great Plains region were productive and accept-able to livestock; it was believed that they provided the best groundcover to protect the mantle of soil and were the most effective plants in utilizing environmental growth factors to fix carbon and to cycle nutrients. Dyksterhuis (1949) regarded the relative carrying capacity of the range as a reflection of the proportion of present vegetation which was original vegetation for that site, although his 1952 paper conceded that a departure of 25% from the climax still indicated range in excellent condition. Heady (1960) referred to 'tall and coarse grasses of low value for livestock use' in East Africa, and suggested that 'if "climax" vegetation need no longer be the ultimate goal in range improvement' range evaluation should continue to study plant succession to identify conditions favouring preferred grass-land stages. 'Productivity' in the range management context usually implies an output of animal product, and the occurrence of unpalatable perennial bunch grasses in relatively undisturbed situations modifies the value accorded the climax.

Predominance of abiotic factors

The emphasis given to grazing pressure as a dominant control of succes-sional processes is inappropriate in some environments, since abiotic factors may exert the primary control of plant dynamics. In western Queensland Roe (1987) noted a gap of 42 years between successive strong pulses of recruitment of *Astrebla lappacea*, indicating that grazing manage-ment which influenced the dynamics of this plant would need an episodic and opportunistic character. O'Connor (1985, cited by Walker, 1988) reviewed 72 long-term grazing experiments in southern Africa; he con-cluded that 'the effects of climate overrode those of grazing treatments'.

Stability and resilience

Teaching of range management has emphasized the quest for an equilib-rium condition, attained by choosing a SR 'which establishes a long-term balance between the pressure of grazing and the successional tendency' (Westoby *et al.*, 1989). A stable equilibrium in vegetation in which a system returns to an equilibrium state after a temporary disturbance is especially difficult to sustain in semi-arid environments or in environments with a high variability of rainfall. Stability has been defined in the different terms of plant species composition or of plant production, which implies continu-ity of forage supply (Walker, 1993).

Much has been made of plant biodiversity as contributing to the maintenance of system structure, the efficiency of resource utilization (Hadley, 1993) and the stability of plant production. McNaughton's (1985) study of vegetation production under grazing at Serengeti, Tanzania is widely quoted to indicate the value of species diversity in contributing to production stability, but diversity of plant species only accounted for 15–16% of the variance in ameliorating biomass instability.

Resilience in ecological systems may be a more realistic goal; Holling (1973) defines this as 'a measure of the persistence of systems and of their ability to absorb change and disturbance and still maintain the same relationships between populations or state variables'. Walker *et al.* (1981) regard resilience as the capacity to adapt to change by exploiting instabilities in contrast to the capacity to absorb disturbance by subsequent reversion to a steady equilibrium state.

Multiple states

Holling (1973) refers to the lack of homogeneity in the natural world and the presence of a mosaic of spatial elements with distinct characteristics linked by mechanisms of biological and physical transport. The linear Clementsian model of succession does not fit the inherent coexistence of several domains of attraction, which many grassland scientists now regard as being better encompassed by the 'state-and-transition' model in which the dynamics of grassland can be described by a set of discrete 'states' of the vegetation and a set of discrete 'transitions' between states (Westoby *et al.*, 1989). This also accommodates better the invasion of grassland by woody weeds or exotic herbaceous species. At the Second International Rangeland Congress various speakers spoke of their dissatisfaction with the Clementsian model but regretted the absence of an alternative paradigm (for example, Foin, 1986). Since then the utility of the 'state-and-transition' model has been increasingly recognized.

5.2.2 Development of the state-and-transition model

The state-and-transition model when applied to rangeland requires first that the various stable states at a vegetation site be identified and catalogued; this requires botanical survey and analysis, preferably using ordination techniques, which is aided by understanding of subsite history and by ecological knowledge of the plant species and their environmental adaptation. The second requirement is to catalogue the transitions between the states which are known to occur and to assess the causes of the transition:

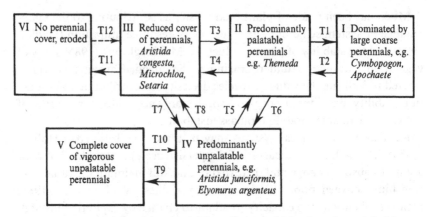

Figure 5.3 State-and-transition model for tall grassveld (both moist and dry) in South Africa. (From Westoby *et al.*, 1989.)

natural 'events' such as weather, fire, flooding or management 'actions' such as a change in SR, fertilizer application, and reseeding (Westoby *et al.*, 1989). The third requirement is to develop a more sophisticated understanding of the dynamics of transitions, so that thresholds for change (Friedel, 1991) may be quantified and the interacting components and processes and their probability of occurrence may be modelled in a pragmatic and meaningful way.

Tall grassveld in South Africa

Illustrations of state-and-transition models are taken from contrasting vegetation types. The first (Westoby *et al.*, 1989) is tall grassveld (both moist and dry) in South Africa and represents a more generalized version than that of the Transvaal highveld described in successional terms in Table 5.1; it is based on Tainton (1981) and observations by B.H. Walker.

State I (Figure 5.3) is dominated by large, coarse perennial grasses ('increaser 1' species).

State II has palatable perennial grasses such as *Themeda triandra*, *Eragrostis racemosa* and *E. capensis* ('decreaser' species), together with some increaser species.

State III is dominated by 'increaser 2' species such as *Aristida congesta*, *Microchloa caffra* and *Setaria flabellata*, but has substantial bare ground, annuals and short-grass perennials.

State IV has large, established tufts of unpalatable grasses such as *Aristida junciformis* and *Elyonurus argenteus* ('increaser 3' species) with little bare ground and some decreaser and increaser 2 species.

State V has a vigorous full cover of increaser 3 species.
State VI has bare ground and annual plants.

The catalogue of transitions, whose driving forces are often speculative, is as follows:

Transition 1. (T1 in Figure 5.3). This may be due to extended rest from grazing; alternatively State I is an edaphic variant of State II or may be grouped with State IV.

Transition 2. Lenient grazing.

Transition 3. Complete or nearly complete relaxation of grazing pressure.

Transition 4. Moderate to heavy grazing imposed so that stock consume unpalatable species.

Transition 5. Hypotheses: (i) it does not occur; (ii) long-term total destocking; (iii) opportunistic timing of grazing at the beginning of particular wet seasons according to whether the timing of rains favours the growth of palatable species, indicating lenient grazing at that stage.

Transition 6. Hypotheses: (i) early season grazing in particular seasons when the rainfall pattern disadvantages the growth of palatable species; (ii) moderate grazing leading to selective consumption of palatable species.

Transition 7. Relaxed grazing pressure (less relaxation than T3) but selective consumption of palatable species.

Transition 8. Very heavy short-duration grazings or fire, manipulated to favour regeneration of palatable perennials.

Transition 9. As for T6, continued to the complete dominance of unpalatable perennials.

Transition 10. May not be feasible, due to complete occupation by unpalatable perennials; alternatively very slow change.

Transition 11. Continued heavy grazing leading to tussock extinction, loss of seed bank, and severe soil erosion.

Transition 12. Reseeding and soil reclamation.

Westoby *et al.* (1989) summarize two routes by which the productive capacity of the tall grassveld may be reduced: (i) overgrazing down from State II to State III and eventually to State VI destroys the resource; alternatively, (ii) State V may be the endpoint, a more significant hazard since T6 occurs at moderate SR and recovery from State IV is much slower than from State III.

The three most common types of rangeland degradation are (i) the shift in botanical composition from perennial grasses to annuals and bare ground, leading to soil erosion; (ii) the change to perennial grasses which are unacceptable to stock; and (iii) the invasion or increase of woody shrubs.

Sagebrush–grass in the Great Basin region

The second example (Laycock, 1991) deals with the last of these and describes the sagebrush (predominantly *Artemisia tridentata*) – grass type which occupies *c.* 50 M ha in the Great Basin region and adjacent areas of USA. The introduction of large numbers of ruminants in the late nineteenth century caused a shift towards sagebrush dominance and a reduced contribution from productive understory grasses and forbs. Fire reduced sagebrush, when the fuel load was opportune, but resting from grazing has had mixed success in restoring the herbaceous component.

The vegetation states are shown in Figure 5.4 and the catalogue of transitions is as follows:

Transition 1.   (T1 in Figure 5.4). Heavy continued grazing and rainfall conducive for seedling establishment of sagebrush.

Transition 2.   Usually inapplicable; route through T3 and T5 more common.

Transition 3.   Fire killing sagebrush; alternatively, biological agents such as insects, ungulates (such as deer or sheep) defoliating over a long period.

Transition 4.   Uncontrolled heavy grazing and reduced vigour of perennial grasses.

Transition 5.   Light grazing.

A favourable climate for annuals such as cheat grass (*Bromus tectorum*) may favour the following:

Transition 6.   Annual grasses replace perennial grasses under continued heavy grazing.

Transition 7.   Usually inapplicable; unlikely if annual grasses well adapted.

Transition 8.   Adult sagebrush destroyed by burning but sagebrush seed bank present.

Transition 9.   Absence of repeated fires permit sagebrush maturity.

Transition 10. Repeated burns kill sagebrush seedlings and deplete seed bank.

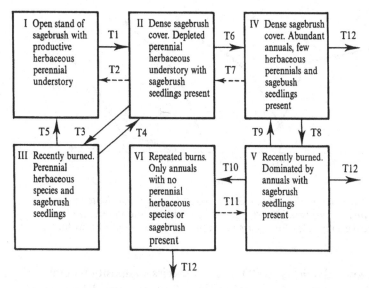

Figure 5.4. State-and-transition model for a sagebrush (*Artemisia tridentata*) – grass ecosystem in the Great Basin region of USA. (From Laycock, 1991.)

Transition 11. Usually inapplicable; requires sagebrush seed bank.
Transition 12. Reseeding of adapted perennials.

Fire-sensitive woody shrubs in many regions of the world display similar patterns of response, and the concatenation of circumstances which can promote an adequate herbaceous fuel load together with subsequent grass-land regeneration, associated with the recognition by the manager of management opportunities, are required for shrub control.

There are now many scientists devising state-and-transition models to describe the multiple grassland states which occur in particular districts and investigating the driving processes of transitions (for example, McIvor & Scanlan, 1994), and the definition of their end points. Jones (1992) reported the recovery from grazing of *Setaria sphacelata* var. *sericea* pasture, planted with the twining legume *Macroptilium atropurpureum,* at Samford, Queensland (lat. 27°, 1100 mm rainfall). This perennial pasture (State I) was overgrazed at 3 b ha^{-1} from 1969 to 1973 and at 2 b ha^{-1} there-after. The sod-forming grasses *Digitaria didactyla* and *Axonopus affinis* invaded, became dominant by 1977, and constituted State II pasture. Exclosures were installed in the pasture at intervals of two years from 1973 to 1983, and yield measured for four subsequent years until 1985. Figure 5.5 shows the steep decrease in State I after 1976. The pasture rested from

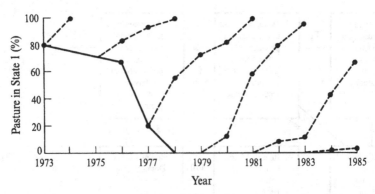

Figure 5.5. Proportion of pasture yield in State I (*Setaria sphacelata* var. *sericea* – *Macroptilium atropurpureum*) under grazing (solid line) and in ungrazed areas (broken lines) exclosed for four years in south east Queensland. (From Jones, 1992.)

grazing recovered quickly to 100% State I species composition in the early phases of degradation, but the transition back to State I became progressively slower. By 1983 *S. sphacelata* var. *sericea* was virtually extinct, with no seed reserves, whilst *M. atropurpureum*, which produces hard seeds, was reduced to 23 seeds m^{-2}; visual observations subsequent to 1985 showed no recovery to State I and the transition phase to a stable, irreversible State II pasture was complete by 1983.

5.3 Dynamics of botanical change

The processes which determine botanical change can be understood through different emphases. Many grassland scientists focus on the impact of grazing animals and of cutting management; plant response to defoliation was discussed in chapter 4, whilst selective grazing and manipulation of grazing pressure is treated in chapter 6. These biotic factors operate within a wider system in which the nature of plant adaptation to the local environment modifies their impact and in which other management interventions (fire, fertilizer application, reseeding, herbicides) may override them. In some environments abiotic factors such as drought or the abnormal occurrence of cold may be the events which alter the composition of the plant population and so change the predominance of particular species and the structure of vegetation.

5.3.1 The competition–stress–disturbance model

The characterization of the environmental adaptation of planted tropical pastures has been attempted (Humphreys, 1981) in terms of comparative resistance to climatic, edaphic or biotic stresses and of the capacity of plants to interfere (Harper, 1977) with the availability of environmental growth factors to their neighbours. The understanding of the environmental adaptation of temperate grasslands, as in the UK, has advanced further with Grime's (1979) conceptual model which adds disturbance to the competition–stress nexus, and distinguishes plant behaviour in the established and regenerative grassland phases.

In established grassland Grime (1979) characterizes the adaptive features of plants according to their predominant strategy for ecological success:

(i) Competitors. These plants exploit conditions of low stress and low disturbance. He defines competition as 'the tendency of neighbouring plants to utilize the same quantum of light, ion of a mineral nutrient, molecule of water, or volume of space'; this definition refers exclusively to plant capture of resources and does not include other mechanisms by which the environment of a neighbouring plant may be made less amenable to it.

(ii) Stress-tolerators. These plants are successful in high stress – low disturbance situations, and stress is defined as 'the external constraints which limit the rate of dry matter production of all or part of the vegetation'. Many scientists have reserved the wider concept of resistance as a generic term to embrace escape, avoidance of stress, and tolerance of stress actually imposed on the plant (Levitt, 1972; Ludlow, 1980). Tilman (1988) defines competition in a sense which emphasizes competitive success as related to plants having a low equilibrium resource requirement, which impinges on the concept of stress tolerance; Grace (1991) has clarified these semantic differences.

(iii) Ruderals. These plants are adapted to low stress – high disturbance situations. Bare ground, caused by retiring land from cultivation, fire, trampling or mowing, may not be associated with low fertility, despite the low density of vegetation. Grime (1979) regards disturbance as consisting of 'the mechanisms which limit the plant biomass by causing its partial or total destruction'.

The competition–stress–disturbance (C–S–D or C–S–R) model provides a simplistic framework in which plants may be placed on gradients which

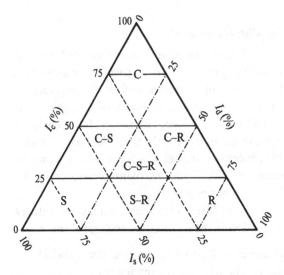

Figure 5.6. Model describing the various equilibria between competition,
C (I_C ——), stress, S (I_S — —) and disturbance, D (I_D — · —) as strategies of
ecological success. (From Grime, 1977.)

then reflect their adaptive function. Figure 5.6 describes the various equilibria between competition, stress and disturbance in established vegetation
(Grime, 1977). Since most grasslands are continually grazed, adapted
species adjust to at least minor forms of disturbance, and most species will
be located away from the apex of the diagram (Grime, Hodgson & Hunt,
1988). The main combinations with European herbage examples are (i) the
competitive ruderals (C–R) adapted to low stress and competition
restricted by disturbance, such as *Hordeum murinum*, *Bromus sterilis* and
Lolium multiflorum; (ii) stress-tolerant ruderals (S–R) in lightly disturbed,
unproductive habitats, such as *Anemone nemorosa* and *Primula veris*, (iii)
stress tolerant competitors (C–S) in relatively undisturbed conditions, such
as *Festuca rubra* and *F. arundinacea*; and (iv) C–S–R strategists, in which
competition is restricted by both stress and disturbance, such as *Festuca
ovina* and *Nardus stricta*.

Grime (1979) identifies plant characteristics associated with each strategy. For example, competitors have a high, dense canopy of leaves with
extensive lateral spread, relatively short longevity of leaves, and peak leaf
production coinciding with periods of maximum potential productivity,
whilst ruderals usually have small stature, limited lateral canopy spread,
and a short phase of leaf production, and stress-tolerators have intermediate characteristics. Ruderals invest more resources in flowering and produc-

tion of seed, especially dormant seed, stress-tolerator plants have well-developed storage systems, and competitors incorporate minerals and assimilate into vegetative structures.

The plant regenerative strategies which are reflected in botanical change are categorized as follows (Grime, 1979):

(i) Vegetative expansion. This exhibits relatively low risk of mortality to offspring since there is transport between parent and offspring and is most common in perennials in undisturbed habitats.
(ii) Seasonal regeneration in vegetation gaps. Seasonal occurrence of vegetation gaps provides opportunity for the colonization of bare ground or sparse vegetation by seed or, less usually, vegetative propagules (Picket & White, 1985).
(iii) Regeneration from a persistent seed bank. The transient seed banks of *Hordeum murinum* and *Lolium perenne*, renewed annually, might be contrasted with the large, persistent seed banks of *Stellaria medea, Calluna vulgaris* (Grime, 1979) and the herbaceous legumes.
(iv) Regeneration involving widely dispersed seeds. This is associated with habitats subject to spatially unpredictable disturbance (McIvor, 1993) and seeds adapted for transport by either wind or water.
(v) Regeneration from a bank of persistent seedlings. Seedlings of trees and shrubs are often present in an understorey, suppressed by shade or other competitive factors until death of mature trees or disturbance creates opportunity for faster growth.

An understanding of the relative significance of these regenerative mechanisms in particular grassland situations (Fenner, 1992) can guide management and also contribute to identifying criteria of merit in plant improvement programmes, as has occurred, for example, in the management of seed banks of annual *Medicago* spp. and *Trifolium subterraneum* in seasonally wet/dry climates (Carter and Porter, 1993).

5.3.2 Population dynamics

Botanical change is underpinned by the insights of population biology (Harper, 1977; Freysen & Woldendorp, 1978; Begon, Harper & Townsend, 1986). J.L. Harper was especially influential in developing concepts of pasture dynamics; a central thesis (Harper, 1978) was that 'Fitness does not depend on maximizing physiologic function but in leaving *more* plant descendants than neighbouring plants are able to do'. Shifts in plant dominance reflect changes in the density of individual plants or tillers and the

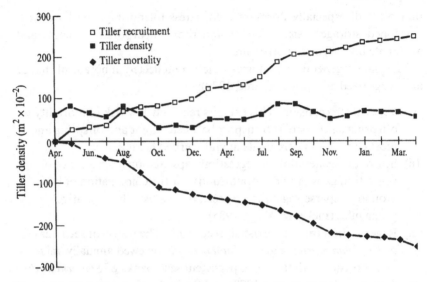

Figure 5.7. Live tiller density, recruitment and mortality in *Lolium perenne* grazed by sheep at Little Wittenham, UK. (From Briske & Silvertown, 1993.)

different success of plants in replacing themselves and making incursions into adjacent areas.

Tiller dynamics

In benign grassland environments in which drastic disturbance is infrequent, vegetative growth is more consistent than growth from seed, because of the import of resources by younger tillers (Briske & Silvertown, 1993). The Grassland Institute at Hurley, UK was especially influential in developing the concept of the grass sward as a dynamic community of tillers whose net productivity reflects the changing pattern of the integrated life histories of the component tillers (Langer, 1963); there were many detailed studies of the seasonal and management factors which influence the growth and longevity of tillers.

Tiller populations exhibit considerable homeostasis in their dynamics. Compensatory effects between leaf number and leaf size, and between shoot weight and shoot density (as in *Cenchrus ciliaris* and *Panicum maximum* var. *trichoglume*, Humphreys & Robinson, 1966) cushion the influence of management variations on net productivity. Similarly, tiller recruitment may balance tiller mortality to give considerable stability in tiller density. This is exemplified in Figure 5.7, which shows a relatively constant live tiller density in a *Lolium perenne* pasture grazed by sheep near Oxford, UK, despite the mortality of more than 20 000 tillers m^{-2} over a two-year measurement period. Only *c.* 200 seedlings were recorded, and plant replacement from seed was negligible in this pasture.

Seedling dynamics

Studies of the life histories of individual plants have revealed to grassland scientists the short-lived character of many 'perennial' species, and the need to organize management which facilitates flowering and seed production, accretion to soil seed reserves, and successful seedling regeneration and survival to complete the flowering cycle upon which plant replacement depends. The greater longevity of shrub legumes than that of herbage legumes simplifies management of the former. The influence of environment on the pathways of persistence of *Trifolium repens* provides an instructive example. I noted a grassland scientist in the UK receiving an accolade for having observed the rare event of a *T. repens* seedling in a grazed pasture, whilst in the subtropics *T. repens* is treated as an annual or weak perennial; near Grafton, New South Wales (lat. 30° S) management is directed to late spring seed production followed by successful seedling regeneration in February, March or April (D.L. Garden, unpublished data, illustrated as Figure 5.8 in Humphreys, 1991). For tropical herbage legumes the half-life under grazing may vary from *c.* 3 months for *Stylosanthes hamata* (Gardener, 1982) to 24 months for *Desmodium intortum* (Humphreys, 1991).

Longevity and inertia

Botanical change may be observed in synchrony with climatic or management changes. In some communities there is a strong lag in the association, and only long-term climatic trends eventually have an impact (Archer, 1993b). Vegetation established under one set of conditions may persist in altered conditions since the original plants have high longevity, or alternatively an established long-term seed bank may provide infrequent but sufficient opportunity for plant replacement. The effectiveness of particular events in driving botanical change may depend upon the age structure of the plant population, which affects its vulnerability to incursion. These changes are most probable on the boundaries of plant communities, which are the most likely sites for species retreat or advancement, although satellite stands can also be effective sources of weed invasion (Moody & Mack, 1988).

Pulses of recruitment

Ecologists working in semi-arid regions (Noy-Meir, 1980) have helped grassland scientists to an understanding of the discontinuous nature of much botanical change. In areas of erratic rainfall pulses of plant recruitment are driven by events of unusual occurrence, often involving the concatenation of differing circumstances: disturbance caused by fire or drought creating a gap coincident with viable soil seed reserves and the occurrence of good germination and follow-up rains, or the unusual accumulation of a herbaceous fuel load, and a fire destroying fire-sensitive woody plants at a time when the seed reserves of the latter are depleted. These instances have made grassland scientists in humid areas more aware of event-driven changes, perhaps associated with some other sporadic occurrence such as a transitory disease epidemic, unusually cloudy conditions affecting seed-setting which has an influence on plant dominance. Each species in a community may be constrained by a different combination of resources (Chapin III *et al.*, 1987) whose pattern of availability may create a complex mosaic. This predicates an opportunistic decision-making framework for grassland management, as modified by the knowledge gained from previous local decisions (Danckwerts, O'Reagain & O'Connor, 1993).

5.4 Assessment of land condition

5.4.1 The development of condition indices

The assessment of 'range' or land condition is based on many different regional systems and continues to evolve in divergent directions. Some of the early definitions required that the observer be within the particular institutional culture which emphasized, for example, the objective of a return to the climax, if the pietistic content of a statement were to be understood. Thus, L. Ellison in 1951 stated that 'condition is the character of the vegetal cover and the soil, under man's use, in relation to what *it ought to be*' (Ellison, Croft & Bailey, 1951, my italics). A basic thrust in range management has been the comparison of the current site condition with its potential condition. This is assessed through the interacting components of production, botanical composition, cover and soil erosion and their 'trend', which then provides a management guide concerning the appropriateness or otherwise of current stocking policy.

Dyksterhuis (1949, 1952, 1958) developed the Quantitative Climax

Index, which was widely used in the USA and Canada (Coupland, Skoglund & Heard, 1960); this provided a measured distance along a successional gradient. The definition of indicator plants and of 'key species' assisted rangeland managers in identifying the position of the site on the gradient and it was believed that botanical composition and productivity were closely linked.

Each range type has differing requirements for the assessment of degradation (Wilson & Tupper, 1982) and differing regional schemes of monitoring range condition have evolved (for example Ellison, Croft & Bailey, 1951; Roberts, 1970; Lendon, Lamacraft & Osmond, 1976; Foran, Tainton & Booysen, 1978; Foran, Bastin & Shaw, 1986; Bastin *et al.*, 1993). Burnside & Faithfull (1993) have drawn attention to the relatively low level of reliability and consistency in judgement made by pastoralists about range trend, and the need for intensive training if range condition assessments are to be meaningful. A recent integrative concept with merit is that 'rangeland degradation is a measure of the reduction in the capacity of a landscape to produce forage from rainfall' (Walker, 1993); this requires the development of new benchmarks.

A study from tropical tall grass pasture lands in Australia (McIvor, Ash & Cook, 1995) is chosen for illustration since animal production is also available and is presented later in the chapter. The condition index was the product of vegetation and soil components:

(i) *Vegetation*. This was derived from (i) species composition, categorized as desirable for legumes and perennial grasses (except *Aristida* spp.), a lower weighting for annual grasses, and zero weighting for other species (including woody shrubs as undesirable); and (ii) productive capacity, based on a linear equation using basal area as the variable.

(ii) *Soil*. The soil component was the sum of (i) surface crust development, weighted as zero with surface crusted with evidence of dispersed clay, since infiltration rate was reduced; (ii) soil surface microrelief, rough surfaces providing better water retention; (iii) litter and cryptogram cover; and (iv) erosion features, including presence of rills, pedestals under grass plants, evidence of soil movement, and sheet erosion.

Ten experimental sites with plots covering a range of conditions were established in northern Australia. Condition indices were generally well related to soil chemical and physical attributes. Relative yield was measured by referring plot yields to the maximum yield (= 100) at each site, and significant linear relationships between relative yields and condition indices occurred at all sites. Most perennial grasses were more abundant and annual grasses less abundant on high-condition plots, as illustrated subsequently in Table 5.2.

The condition index developed by McIvor *et al.* (1995) was clearly successful in integrating vegetation and soil components to predict herbage growth. However, further work on plucked and total herbage samples with respect to the nutrient concentration, nutrient uptake and digestibility gave divergent perceptions (Ash & McIvor, 1995). Uptake of N and P, and P concentration, decreased as land condition declined. The reverse was true for N concentration and digestibility. These indicators of forage quality significantly increased over all sites as land condition declined. This may be due simply to N dilution where better herbage production occurred, and to arrested growth on low-condition plots leading to lesser accumulation of cellulose, hemicellulose and lignin; nevertheless, these findings complicate the application of the condition index to decision making in management.

5.4.2 Animal production and land condition

The scarcity of attempts to relate animal production to land or range condition represents a great weakness in grassland research; scientists have often assumed a connection between animal output and dry matter production and soil cover or relied on intuitive perceptions. As mentioned earlier, it was fortunate that the climax perennial grasses apparently reflected the highest potential for animal production in the mid-west of northern America where successional theory developed. The limited number of studies in other regions produced unexpected results, which place conventional concepts of range condition under great challenge.

The then heretical R.L. Davidson (1964) regarded the early seral stages of the Transvaal highveld, South Africa, dominated by *Eragrostis curvula* and *Cynodon dactylon*, as more productive, more responsive to N fertilizer and less susceptible to soil erosion than climax grassland, dominated by the tall bunch grasses *Trachypogon spicatus*, *Elyonurus argenteus* and *Tristachya hispida*. At Frankenwald (lat. 26° S, 760 mm annual rainfall, 1500 mm altitude) Gillard (1966) showed that LWG was 20% greater on seral grassland than on climax grassland when these were grazed at the same SR of 2.3 steers ha^{-1}; average N% of herbage was respectively 1.55 and 1.23. However, in this study differences in botanical composition were confounded with a differing earlier history of fertilizer application, although similar changes in botanical composition might be induced by differing SR alone, as mentioned earlier (Gillard, 1969). A further African pointer (Harrington & Pratchett, 1974a) was an SR experiment at Ankole, Uganda (890 mm annual rainfall) on natural grassland dominated by *Themeda triandra, Hyparrhenia filipendula, Digitaria maitlandii* and *Brachiaria decumbens*. Marked shifts in botanical composition occurred over four years. Cattle growth decreased with increasing SR, but there was an interesting departure from the linearity expected between LWG and SR (Jones & Sandland, 1974). The highest SR of 1.7 b ha^{-1} gave higher LWG than anticipated and the authors ascribed this to the greater content in this treatment of the elite native *B. decumbens*, which exhibited a higher protein content (Harrington & Pratchett, 1974b).

The performance of weaner steers stocked continuously at a lenient level of utilization, which displays the potential for individual LWG since maximum opportunity for diet selection applies, and at seasonally adjusted SRs to give similar levels of utilization, was studied on natural grassland whose condition had been altered by differing histories of grazing (Ash *et al.*, 1996). This study was carried out in tropical savanna near Charters Towers, Queensland (lat. 19° S, 535 mm annual rainfall, 300 m altitude); similar results were obtained at Katherine, Northern Territory and data from the former location are summarized here. The botanical composition of the areas which had been previously lightly (State I) or heavily (State II) grazed (Table 5.2) was categorized according to the content of perennial grasses decreasing or increasing with grazing, of annual grasses, and of native legumes and forbs. Samples of these plucked towards the end of the growing season showed a higher N% in annual than in perennial grasses, and considerably higher N% in native legumes and forbs. Both states exhibited similar groundcover at the inception of the experiment, which was limited to two years so that the results would reflect current productivity of the two states.

A year-round continuous stocking treatment was imposed at 0.4 b ha^{-1}, estimated to give 5 and 9% utilization in States I and II respectively; initial annual dry matter production was 3540 and 2060 kg ha^{-1} in the two states. Additionally, SRs for the early and late wet, and early and late dry seasons were adjusted to give low (20%), medium (35%) and high (59%) levels of proportional utilization. Steer LWG (Figure 5.8a) was substantially greater on the State II 'degraded' grassland under continuous stocking and at the low level of utilization; these differences were greatest in the early wet season and least in the late dry season. The actual seasonal SRs imposed were necessarily greater on the higher yielding State I land than on State II land (Figure 5.8b). On both states there was a significant negative linear relationship between LWG and SR, but the intercepts and slopes of the

Table 5.2 *Botanical composition and N percentage of plucked samples for two condition states at Charters Towers, North Queensland*

		Composition (%)	
Group	N (%)	State I	State II
Decreaser perennial grasses	1.31[b]	83	35
Heteropogon contortus			
Bothriochloa ewartiana			
Dichanthium sericeum			
D. fecundum			
Themeda triandra			
Increaser perennial grasses	1.31[b]	7	19
Bothriochloa decipiens			
Aristida calycina			
Chrysopogon fallax			
Bothriochloa pertusa[a]			
Urochloa mosambicensis[a]			
Annual grasses	1.40	5	27
Urochloa panicoides[a]			
Sporobolus australasicus			
Tragus australiana			
Native legumes and forbs	2.40[b]	5	19
Indigofera colutea			
Vigna lanceolata			
Pterocaulen redolens			
Sida spinosa			

Note:
[a] Introduced species; [b] mean of both classes.
Source: Ash *et al.* (1996).

regression lines were different, indicating the superior LWG of State II land at low SR, its inferior performance at high SR, and the greater stability of production of State I land over a range of SRs.

Seasonal differences in LWG were well related to the percentage of green leaf on offer, but the same green leaf percentage resulted in greater LWG in State II compared with State I. N concentrations in forage and in faeces were significantly greater in State II than in State I in the early wet season, but P concentration, which also decreased with advancing season, did not differ between the two states. The relative proportions of C_3 and C_4 species in the diet, as determined from analysis of ^{13}C in faeces, showed greatest content of C_3 species in the late wet; dietary content of C_3 species was 8–20% greater in State II than in State I grassland.

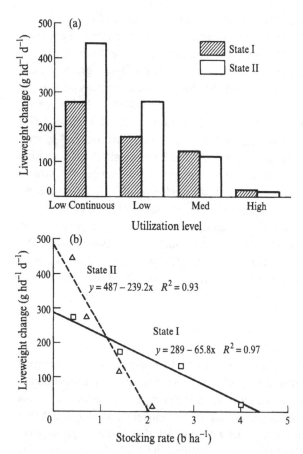

Figure 5.8. Relationships between liveweight change and (a) level of utilization and (b) stocking rate for two pasture states in tropical tall grass near Charters Towers, Queensland. (From Ash *et al.*, 1996.)

This seminal study indicates the need to assess the effects of changes in botanical composition in terms of animal response, and the importance of the SR used in the assessment. 'Regression' at this location to increaser perennial grasses, annual grasses, and incidentally to native legumes and forbs, led to a higher quality diet for the grazing animal and reflected a homeostasis in the ecosystem where decreasing pasture availability under high SR was compensated by higher pasture quality, unless the forage supply was exhausted. The study was complicated by the invasion of introduced grasses which had become naturalized in northern Australia and which provide good ground cover under heavy grazing. These results may be contrasted with other Australian studies.

A change in botanical composition from a chenopod shrubland to annual grass pasture had negligible effects on animal production (Wilson & Leigh, 1967; Graetz, 1986). Where perennial grassland was replaced by annual grassland in central western New South Wales as a result of heavy grazing, wool growth and sheep body weight benefited, since the annual plants were better adapted to respond to winter rainfall (Robards, Michalk & Pither, 1978). These findings are reinforced by a SR study in Natal (Hatch & Tainton, 1993) in which cattle growth was slightly better at a site with 'poorer' species composition than at a site with 'superior' species composition and in which range condition score at each site did not significantly influence net return.

Relationships between range condition and animal production clearly need to be developed in quantitative terms so that land managers are able to assess the risks inherent in intensifying SR, which also impinges upon the ease of their capacity to vary SR seasonally and between years. The situations described above are inherently different from ecological conditions which lead to the invasion of inedible woody species or grasses unacceptable to stock.

5.5 Conclusion

Many rangeland scientists, having lost their faith in the universal applicability of Clementsian succession, are seeking new certainties and are especially vulnerable to the winds of scientific fashion. The latter half of this chapter indicates that these certainties are not currently available, but that wider conceptual frameworks exist within which the search for good paradigms of grassland dynamics may continue, and which are eventually directed to the development of decision-support systems for grassland management.

(i) The first requirement is deeper understanding of the processes which drive change in grassland communities and which is emerging from rigorous ecophysiological studies which have the necessary autecological bias. This may be directed especially to defining thresholds of change (Friedel, 1991) if state-and-transition models are to become pragmatic; linked to this there is the desirability of identifying various functional groups of plants (Walker, 1993) defined in terms of nutritional value, seasonal growth phenology, and response to drought and to grazing pressure. The identification of these plant groups is linked to describing site potential and the state of particular resources; great advances have been made in the use of satellite imagery to quantify rangeland communities (Pickup, 1989; Tsuiki, Takahashi & Oku, 1993).

(ii) A recurrent theme in the rangeland literature is the requirement for sustainable resource use. A recognition of multiple stable vegetation states at any site widens the scope for opportunistic management which is responsive to abiotic events and which is not bound by ancient lore which avoids fire or which values moderate SR during all circumstances; this relates especially to the management of grasslands invaded by woody species. Managers continue to seek a safe (but perhaps variable) SR which conserves the vegetation resource, utilizes environmental growth factors effectively, holds fertile topsoil *in situ*, and provides continuity of forage supply to grazing animals. The approach of Scanlan *et al.* (1994) in combining factors of grassland production with forage demand in the one model, as discussed in chapter 6, requires elaboration in other regions.

(iii) Research of these types is more effective if conducted in collaboration with landholders who share the ownership of the research. In many countries government institutions fund research and provide institutional orientations which are dislocated from the goals and perceptions of the landholder who lives from the outputs of the rangeland. Outputs should emerge which formulate functions of welfare based on long-term net economic benefit (Walker, 1993) and which include the components of present economic return with the discounted value of the production potential of the vegetation resource and of the herd, maximized in the long term. Hopefully, these functions can provide decision support systems which ensure the survival of the grazier and the herd, sustained by a resilient grassland.

6
Grazing management

6.1 Introduction

The objectives of grazing management are to synchronize the supply of available forage with the demand of the animals grazing the pasture and to maintain the vigour of acceptable pasture. These are effected by choosing a stocking rate (SR, the number of animals carried on the pasture per unit area) which reflects the long-term capacity of the pasture to produce acceptable forage, as modified by the inputs made by the manager which influence pasture growth and quality. A consequential objective is to transfer the availability of forage in time and to give priority feeding to the classes of stock generating the greatest marginal return (McCall & Sheath, 1993). This involves stocking method: the deferment of grazing in order to accumulate surplus feed, and the rotation of stock around different paddocks according to some defined objectives associated with the botanical composition of the pasture or with farm convenience. This may be directed to the provision of a sequence of different forages in order to maintain continuity of forage supply (Humphreys, 1991). The species of animals grazed also impacts on the sustainability of forage resources. Grassland scientists have long been concerned about the effects of overgrazing and the degradation of the landscape, as outlined in chapter 5.

It was not until the 1950s that research on the relationships between SR and animal production started to become fashionable; prior to this a good deal of emphasis was given to stocking method and the need to rest pastures from grazing and avoid the dangerous effects of selective grazing on botanical composition. At the IV International Grassland Congress Ll. Iorweth Jones (1937) stated 'no one is likely to dispute the advantages of rotational grazing . . .'. In his presidential address R.G. Stapledon (1937) advocated 'rotation in time and rotation in space' and dual attention to the needs of the sward and the needs of the animal: 'by adopting a system of

156

rotational grazing – intermittent with proper periods 'on' and proper periods 'off' – the animal can be given somewhere every day what it requires and the swards need never suffer'. At this same Congress Martin Jones (1937) illustrated the benefits of seasonal variation in SR and the conservation of silage on the botanical composition of the sward. There were early sceptics. Donald (1946) stated that in no instance had 'consistent, economically worthwhile differences been established in favour of rotational grazing'. Stocking rate and stocking method and their perception by the grazier form the basic components of grazing systems.

6.2 Stocking rate and animal production

6.2.1 Development of the model

At the VI International Grassland Congress McMeekan (1952) referred to the interdependence of grassland and livestock and placed selection of the number of stock to be carried as the first essential in the matching of animal needs to available nutrients. At the next Congress McMeekan (1956) demonstrated that '. . . even extreme differences in grazing methods are associated with relatively small effects upon efficiency' but that 'Rate of stocking is by far the most powerful weapon . . . influencing efficiency on a per acre basis'. He reported New Zealand experiments which doubled ewe SR, increased lamb and wool output per unit area by 80 and 86% respectively, whilst increasing cow SR by 67% increased production of butter fat per unit area by 42%. McMeekan (1961) continued to expound this emphasis at the VIII International Grassland Congress but presented data which suggested controlled rotational grazing of grass–clover swards gave a higher response to increased SR than occurred under set stocking; this comparison was confounded by differences in the amounts of forage conserved and fed under the two stocking methods.

At this Congress the most notable paper on grazing management was that of G.O. Mott (1961), who presented generalized curves relating SR to animal output per head and per unit area (Figure 6.1); these were based on eight SR experiments in the USA and New Zealand. Product output per animal (Y) followed an exponential relationship with SR. At a very low SR, which provided maximum opportunity for diet selection, Y was 12% higher than the value at 'optimum' grazing pressure, and was given a value of zero at $1.5 \times$ SR at optimum grazing pressure. Output per unit area (Z, the product of SR and Y) increased rapidly as pastures were utilized more, reaching a maximal value just beyond the optimum grazing pressure and

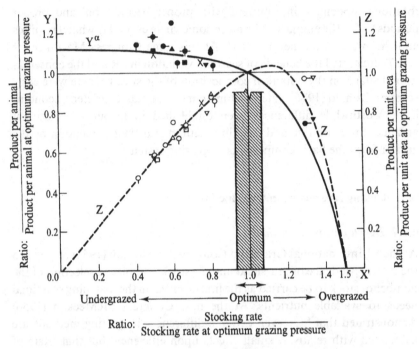

Figure 6.1. Generalized curves indicating the relationship of SR (X') to the ratio product per animal (Y')/product at optimum grazing pressure and the ratio of product per unit area (Z)/ product at optimum grazing pressure. (From Mott, 1961.)

crashing to zero at 1.5 × SR at optimum grazing pressure. The livestock manager was therefore presented with a production model within which choices about SR might be made according to the premium available for high individual animal output, the value of animals and the cost of their maintenance, and the relative value of land.

This was followed by an influential paper by M.E. Riewe (1961) from Texas, which presented a simple linear relationship between SR and output per head, and suggested that the SR at the crash point of no animal gain was not 1.5 × SR at the optimum, as suggested by Mott (1961), but 2 × SR giving maximum gain per unit area. Hildreth & Riewe (1963) provided an economic analysis of this function which accommodated varying purchase and selling prices for fattening steers; the economic optimum SR increased if steer prices rose during the fattening period and vice versa. Jones & Sandland (1974) then tested this model of a linear relationship between SR and LWG per animal over a wide range of grazing experiments with both cattle and sheep on 33 temperate and tropical pastures; the *r* value was 0.99.

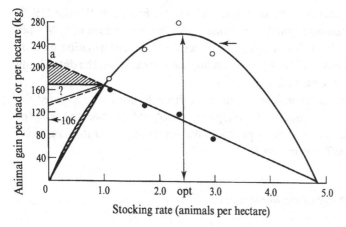

Figure 6.2. Relationship of LWG per animal (●) and per hectare (○) to SR on a pasture at Samford, south east Queensland. (From Jones & Sandland, 1974.)

The Jones & Sandland (1974) model, which is illustrated for one of Jones' experiments with *Setaria sphacelata* var. *sericea* – *Macroptilium atropurpureum* cv. at Samford, Queensland in Figure 6.2, has proved to be remarkably robust and has been applied widely in differing environments with various pastures and livestock outputs. The linear equation $Y = a - bX$, where Y is output per head and X is SR in animals per unit area, becomes a quadratic function $Z = a - bX^2$ when output per unit area (Z) is estimated. Differentiating, SR giving maximum output per unit area is $\frac{a}{2b}$; this occurs at half the potential maximum LWG per head (the intercept a) and zero output occurs at 2 × this SR. In Figure 6.2 the values of a and b are 212 kg and 43.7; maximum LWG hd^{-1} yr^{-1} is 212 kg and each increase in unit SR decreases LWG hd^{-1} by 43.7 kg. Thus, SR giving maximum LWG ha^{-1} is 2.4 b ha^{-1} and maximum LWG is 257 kg ha^{-1}.

This model is at variance with the findings of some other workers who criticized its applicability and mathematics. Harlan (1958) had previously proposed a double exponential function, but Riewe (1961) suggested that Harlan's data give a linear function if SR was expressed as b ha^{-1} and not area per animal. Petersen, Lucas & Mott (1965) added to the confusion with the results of a rat feeding experiment which gave a flat response in LWG hd^{-1} to increased feed availability beyond a critical point and a concave response as the feed supply decreased. This model contained two aberrant assumptions: the amount and character of forage available was independent of SR, and the quality of forage available and of forage consumed was identical; the applicability of the findings to the grazed pasture

160 Grazing management

situation may therefore be discarded. Conniffe, Brown & Walshe (1970) preferred a quadratic function to linearity in the association of LWG per head and SR, whilst Connolly (1976) criticized the transformation used by Jones & Sandland (1974) as one which imposed linearity on the data if only a small part of the curve is considered.

Nevertheless, the experience of the past three decades has validated the utility of the linear model as a management tool, at least over the range of SR encountered in most farm practice (Walker, Hodge & O'Rourke, 1987; Van Heerden & Tainton, 1989).

6.2.2 Applicability of the linear model

Some limitations

The Jones & Sandland model was derived from data of animals grazing pastures year-long or at least for a full grazing season. Instantaneous or short-term responses may be different, especially if compensatory weight gain is involved. Some workers have calculated SR giving maximum LWG ha^{-1}, and arrived at a value extrapolated beyond the SRs actually tested; such a value needs to be treated with extreme caution. Jones & Sandland (1974) use the term 'optimum' SR for the SR giving maximum output per unit area; whilst the use of this designation may be mathematically correct, many producers will not regard animal output per head at one half the production potential as optimal. Several factors determine the optimum SR, not least being the capacity of the pasture to support a SR at that level without deterioration, loss of basal cover, weed ingress, and increased runoff and soil loss. Financial factors usually dictate that the optimum economic SR is less than the SR leading to maximum output per unit area (Carew, 1976; Izak, Anaman & Jones, 1990); the target weight for animal turn-off is influential.

Animal products

The discussion to this point relates to LWG. For wool production the linear model is robust (Lloyd, 1966) and the slope of the regression line is usually less than for LWG, since sheep are compulsive wool growers. Milk production, expressed through the initial growth of young lambs or calves or through milk sold from the farm, tends to be cushioned by changes in the body reserves of the mothers to accommodate variation in forage intake (for example, Leaver, 1982). In Normandy an increase in SR of 30% led to only 5% decrease in the milk output of individual cows (Hoden *et al.*, 1991).

The assessment of SR effects on dairy production has been constrained by the unwillingness of scientists to expose cows to the stress of high SR which would generate sub-maximal output per unit area. Linear functions have been observed (Cowan, Byford & Stobbs, 1975; Davison, Cowan and Shepherd, 1985); the latter authors also suggest a sigmoid relationship between SR and milk cow^{-1} may apply at extreme SRs in which at high SR the mobilization of body reserves for lactation reduces the effect of feed shortage whilst at low SR a plateau appears because of the limited capacity of animals to improve diet quality further through selectivity.

Slope of the regression

The slope of the regression line (the '*b*' value in the equation relating output per head to SR) illustrates the robustness of the pasture in responding to changes in grazing pressure. A low *b* value accommodates more flexible SRs, since variation in SR does not change individual animal output greatly, whilst a high *b* value indicates more stringent control of SR is desirable. Some data from planted tropical pastures (Table 7.1 in Humphreys, 1991) indicate a range in *b* values of 8–57 kg LWG hd^{-1}; that is, for a change in SR of 1 b ha^{-1} LWG hd^{-1} changes by 8–57 kg yr^{-1}. The values for grass pastures fertilized with N are generally lower than for grass–legume pastures. SR is expressed simplistically as b ha^{-1}; LW ha^{-1} has a more precise meaning or may be modified further to metabolic size (e.g. LW$^{0.75}$, Roberts, 1980), whilst the physiological condition of the animals also needs to be taken into account.

Performance at low SR

The intercept '*a*' on Figure 6.2 on the Y axis of the linear equation relating LWG hd^{-1} and SR indicates the quality of the pasture and its potential for individual output. There is some uncertainty, indicated by the dotted lines on the left in Figure 6.2, as to whether LWG hd^{-1} continues to increase as SR decreases to very low levels where animals do not compete for the supply of forage and have maximum opportunity to select a diet of the highest nutritive value available. If this is not the case the values for LWG hd^{-1} plateau at low SR; alternatively LWG hd^{-1} may actually decrease. This arises in two circumstances. Botanical composition often alters at differing SR, as discussed in chapter 5; unfavourable changes may occur at very low SR. For example the component of *Trifolium repens* may be lost from *Paspalum dilatatum* pastures at low SR, due to shading, to the disadvantage of nutritive value. The favourable microclimate provided by low SR for the development of the pathogen *Rhizoctonia solani* on the legume *M. atrop-*

urpureum cv. Siratro may decrease the content of this legume in a mixed pasture, giving reduced LWG ha^{-1} at low SR (Walker, 1980). The second circumstance arises where the quality of the diet ingested decreases at low SR for structural reasons, associated either with the accessibility of green leaf material (Vieira, 1985) or with the spatial heterogeneity of grazing and the presence of large ungrazed patches in the pasture during the main growing season (Edye, Williams & Winter, 1978).

Response to inputs

The SR production model has to take into account variation in inputs provided by the manager which influence either the supply of forage or the dependence of the animal on grazed pasture. The optimum SR increases as the productivity of the pasture is enhanced by the replacement of indigenous species with superior exotics, by the control of weeds and the removal of inedible woody shrubs, by supplementary irrigation and by fertilizer application, and additional animals are needed to convert the augmented forage supply to animal product and to offset the cost of the additional inputs. The SR giving maximum LWG ha^{-1} may be increased by a factor of nearly three by increasing inputs of N fertilizer (Mears & Humphreys, 1974b; CIAT, 1976), whilst the application of superphosphate to a grazing resistant legume may so increase soil fertility that *b* values are greatly diminished (Shaw, 1978). Alternatively, the dependence of milk output cow^{-1} on SR decreases as the level of concentrate feeding to dairy cows increases, substitutionary feeding occurs, and pasture availability increases unless SR is also increased (Cowan, Davison & O'Grady, 1977).

Climatic variation

The predictive value of the SR production model is lessened by the year-to-year variation in pasture growth, associated with climatic variability or other causes. Within years (or seasons) there is also an instantaneous variation in the response to SR, according to growing conditions; the *b* value for unit change in SR for yield of milk cow^{-1} d^{-1} at Atherton, Queensland increased from 1 to 5 kg as the growing season advanced (Cowan, Byford & Stobbs, 1975). More sophisticated approaches to SR – feed availability relationships are therefore indicated.

Decisions about SR may be made on the basis of the assessment of climatic risk and the production potential of the pasture in semi-arid areas where adjustment of SR within seasons is not readily feasible and where there are few alternative sources of feed.

One such approach, (Scanlan *et al.*, 1994) which illustrates the need for local equations, attempts to predict a safe carrying capacity for extensive cattle grazing in north eastern Australia. The first step is to estimate the potential pasture productivity for a particular soil type. Many workers have applied a linear relationship with rainfall as rainfall use efficiency (RUE units), which ranges from 2 to 7 kg DM ha^{-1} (mm transpired^{-1}). Variation in this value for a site from year to year (McKeon *et al.*, 1990) is related to daytime vapour pressure deficit (VPD). The pasture production in this region is well related to summer rainfall and a value is stated which is equalled or exceeded in 70% of years. Other levels of risk may be adopted. Thus, growth potential may be estimated as the product of these three components.

Actual pasture growth is then regarded as proportional to (1) pasture basal area (0–5%), leading to a pasture condition index, and (2) interference from the presence of woody overstory, proportional to tree basal area, giving a tree index. The safe level of utilization is regarded by Scanlan *et al.* (1994) as 30% of summer growth and this figure divided by estimated summer consumption gives safe SR.

Under this scheme utilization of less than 30% would occur in at least 70% of years, and this level is considered adequate for the maintenance of grass basal area. This model gave good agreement with graziers' estimates of the safe carrying capacity of their properties but was poorly related to the institutional estimates which had evolved over time. Scanlan *et al.* (1994) suggest the model would be further refined by attention to (1) spatial variability in resource use by grazing animals; (2) more exact definition of the effects of woody plants and of pasture condition on production; (3) herbage consumed by feral and native herbivores; (4) the variation in safe levels of utilization according to pasture type; and (5) more sophisticated assessment of potential pasture productivity. Models for cattle producers in the USA have also been developed.

An SR which is altered annually or at the end of a season can only effect a crude synchrony between forage availability and forage demand (Hart & Hanson, 1993). Variation in the time of mating and the manager's policy with respect to animal purchase and sales can mitigate the discontinuities in forage supply associated with climatic variation, and in many farming systems it is feasible to combine these policies with the use of forages of differing seasonal availability and of supplementary feeding to maintain relatively constant grazing pressure. A continual adjustment of SR to forage availability is the paradigm sought within the limits of feasibility. This requires a local understanding of the responses of animals and of pastures to variation in forage allowance, which is the amount of forage available per animal unit.

6.3 Forage allowance

6.3.1 Development of the concept

Scientists have expressed the forage availability/forage demand relationship either in terms of grazing pressure (animals per unit forage) or of forage allowance, often expressed as kg DM 100 kg LW^{-1} d^{-1} or as g OM kg LW^{-1} d^{-1}. This concept makes it possible to budget feed requirements in relation to pasture productivity. An early exponent was J.C. Knott of the Washington Agricultural Experiment Station, who proposed that available forage be expressed as 'standard cow days', which is calculated as yield of total digestible nutrients (TDN) per acre divided by 16 (Knott, Hodgson & Ellington, 1934).

W.M. Willoughby (1959) was one of the first scientists to propound an asymptotic relationship between pasture availability and animal performance; the rate of sheep LWG increased steeply at low levels of herbage presentation yield as herbage availability increased and reached a plateau beyond which no further increase in growth occurred. This type of relationship has been widely reported for herbage intake (Baker, 1978) and for milk, LWG and wool production (for example, Van der Kley, 1956; Schulz *et al.*, 1959; Hull, Meyer & Kromann, 1961; Greenhalgh *et al.*, 1966 and Adjei, Mislevy & Ward, 1980). In the last quoted study, maximum LWG of yearling cattle of 0.6 kg hd^{-1} d^{-1} on *Cynodon* spp. at Ona, Florida was attained by managing forage allowance at 6–8 kg DM 100 kg^{-1} LW d^{-1}. At high levels of forage allowance the efficiency of defoliation, as measured by the ratio of dietary intake to herbage disappearance, decreases due to increasing senescence and to consumption by non-domestic fauna (Stuth, Kirby & Chmielewski, 1981).

At the VI International Grassland Congress Mott & Lucas (1952) suggested that the comparative productivity of pastures should be assessed when each was stocked at its optimum level, using a 'put-and-take' system of management. The 'put-and-take' method of grazing experimentation was widely adopted in the USA and in Latin American countries, where many grassland scientists were mentored by Mott. In this system each pasture treatment was grazed throughout by the same indicator stock, designated 'testers' by R.E. Blaser, and their individual performance in terms of LWG or milk was 'a function of the nutritive value of the forage and the rate of intake' (Mott & Lucas, 1952). The carrying capacity or animal days per unit area represented the sum of the grazing by tester animals and 'put-and-take' animals which were continually added to or subtracted from the pasture in order to maintain a constant level of forage availability. This might be determined through a target grazing height, forage presentation yield, or the residual pasture present when animals were removed to the next paddock in a rotational grazing system. Output per unit area was then estimated as the product of the individual output of the tester animals and the number of stock days per unit area.

This method of assessing pasture productivity has been criticized on two counts:

Lack of pertinence to the farm situation

Livestock production based wholly on production of grazed pasture lacks the flexibility of SR applied in a 'put-and-take' system. The results are only weakly applicable to a farm situation where feed surpluses are stockpiled in situ in the paddock and feed shortages are endemic. On the other hand, the results are applicable in circumstances where pasture surpluses are conserved, where a variable level of supplementary feeding constitutes a major source of animal diet, or where pasture growth rate is modified by fertilizer or irrigation inputs to reduce discontinuity in forage supply. It is also entirely relevant to the management of high quality special purpose pasture. In this case grazing may be integrated with a larger area of a lower quality base pasture which may also be stocked flexibly to absorb the impact of changes in the SR imposed on the high value special pasture according to the perceived management needs of the latter.

Subjectivity

Decisions about adding and subtracting animals from the pasture treatments may be based on the managerial judgement of the operator, or differing criteria for SR adjustment may be applied to pasture treatments of varying composition or height, implying elements of subjectivity. This may in part be overcome by the introduction of precise decision rules and objective measurements of feed availability or utilization.

6.3.2 Composition of forage allowance

Mannetje & Ebersohn (1980) distinguished grazing conditions where herbage consumption may be a relatively fixed proportion of forage allowance, or where this ratio varies seasonally. In the latter case a better relationship of animal performance to forage allowance occurs if this is expressed as green dry matter (GDM) rather than DM or OM (see for example, Watson & Whiteman, 1981); their relationship was further improved if the content of legume, which enhanced animal performance, was also taken into account. Similarly, equivalent levels of milk production on clover based pastures were attained at 70% of the forage allowance allocated for *Lolium multiflorum* pastures (Chopping, Lowe & Clarke, 1983), and this indicates the need for local equations for particular pasture types. The level of utilization may also be adjusted according to whether the rate of consumption is exceeding the current rate of pasture growth or not; pastures with an increasing content of senescing tissue provide a poorer diet than pastures with a lower content of non-green grass (Ebersohn, Moir & Duncalfe, 1985). Forage allowance may be seasonally adjusted and increased as IVD DM decreases (Guerrero *et al.*, 1984) to provide more opportunity for diet selection. Conditioning of pastures in New Zealand by heavy spring grazing, which reduced reproductive activity, produced early summer swards of greater leafiness and digestibility. These gave greater milk production at the same level of forage allowance than swards not subjected to this conditioning (Hoogendorn, *et al.*, 1992); previous grazing and management history may therefore be taken into account with advantage when forage allowance is determined.

6.4 Sward surface height

The perception that forage allowance might be related to the average surface height of the sward led to a significant management advance, since this is easily measured by the farmer. J. Hodgson (1981, 1985) played a key role in developing and applying this concept in intensively managed temperate pastures and Hodgson, Mackie & Parker (1986) recommended grazing rules for farmers which indicated the necessary percentage change in stock density according to the change in the target sward surface height (SSH) over the previous week.

6.4.1 *Animal production and sward surface height*

Milk production

Dairy production responses are illustrated from a study at North Wyke, Devon, UK (Mayne *et al.*, 1987) in which cows grazed from May to August on *Lolium perenne* swards arranged as 24 one-day paddocks and residual sward height, as measured by a rising-plate stick, was adjusted to 50, 60 and 80 mm by a put-and-take system. Forage allowance averaged 43–91 g OM kg LW^{-1} d^{-1} across treatments (Table 6.1) and the severely grazed sward suf-

Table 6.1 *Pasture characteristics and animal production of swards grazed to differing residual sward heights at North Wyke, UK*

Component	Sward surface height (mm)		
	50	60	80
Forage allowance (g OM kg LW^{-1} d^{-1})	43	55	91
Tiller density (June, $\times$ 10^3 m^{-2})	14	13	11
OM digestibility, forage on offer (g kg DM^{-1})	688	684	654
Nitrogen concentration (g kg DM^{-1})	25.0	24.3	22.7
Herbage intake (kg OM cow^{-1} d^{-1})	10.8	11.5	12.0
Milk yield (kg cow^{-1} d^{-1})	13.7	16.0	17.0
Liveweight change (kg cow^{-1} d^{-1})	-0.15	$+0.07$	$+0.03$
Stocking rate (cows ha^{-1})	6.0	5.5	5.0
Milk yield (kg ha^{-1} d^{-1})	83	88	84
UME output (GJ ha^{-1} d^{-1})	0.72	0.77	0.72

Source: Mayne *et al.* (1987).

fered treading damage in the spring. Tiller density was numerically less in the lax swards, which also presented forage of lower digestibility, greater stem content, but higher N concentration; differences in water soluble carbohydrate and fibre components did not reach significance, and herbage intake was reduced on the severely grazed sward.

This reduction was paralleled (Figure 6.3) by decreased milk production per cow, and reduced body condition. However, milk output per unit area was similar between all treatments, and 6% greater in the medium 60 mm treatment than in the 50 mm treatment, whilst utilized metabolizable energy, UME, followed a similar pattern. The residual sward height of 60 mm, which gave an optimal response, would be equivalent to *c.* 80 mm SSH. However, within the grazing group high yielding cows showed depressed milk yields if pastures were grazed below a residual sward height of 80 mm, whereas the output from low yielding cows was only slightly depressed at 50 mm residual height.

Beef production

The effects of SSH on cow/calf performance on swards dominated by *L. perenne* at Shotts, UK (Wright & Whyte, 1989) were similar to the North Wyke results quoted above. Cow LWG over the period May to September gave a quadratic relationship with SSH (Figure 6.4a) with maximal value at *c.* 90 mm. Calf LWG (Figure 6.4b) was less sensitive to SSH during the

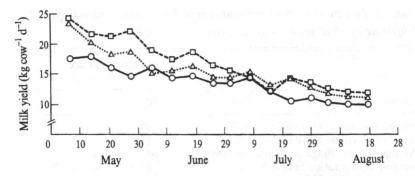

Figure 6.3. Milk yield (kg cow^{-1} d^{-1}) from swards grazed to differing residual heights (50 mm ○, 60 mm △, 80 mm □ at North Wyke, UK. (From Mayne *et al.*, 1987.)

early grazing season from mid-May to mid-August, when cows were suckling, but was increasingly sensitive to SSH as calves became more dependent on herbage intake. At SSH of 110 mm a high proportion of the sward was infrequently grazed and contained much dead material. The sum of cow and calf LWG, expressed as output per unit area, was greatest in the treatment where SSH was *c*. 80–90 mm.

Since the cost of over-wintering feed for beef cows is greater than that of grazed pasture it is economically desirable to maintain SSH at a level favourable for cow body condition, even though calf LWG per unit area during the first half of the growing season is maximized by hard grazing. Wright & Whyte (1989) suggest that grazing which maintains SSH of not more than 80 mm in spring and early summer, which will reduce flowering, followed by SSH of 90–100 mm to favour calf LWG, is desirable. Somewhat similar conclusions were obtained by Morris *et al.* (1993) for LWG of finishing steers and young bulls grazing *L. perenne – T. repens* swards at Palmerston North, New Zealand. Under continuous stocking desirable SSH was *c*. 80–100 mm in the spring, and *c*. 120–150 mm in the autumn in order to maximize individual animal LWG, which was linearly related to intake.

Sheep production

Lamb and meat production on intensively managed temperate pastures may benefit from the management of SSH; this technique is less applicable to extensive wool production. Sheep prehend herbage growing closer to ground level than cattle do, and recommended SSH is relatively less. Lamb production from *L. perenne* swards either with a small clover component or

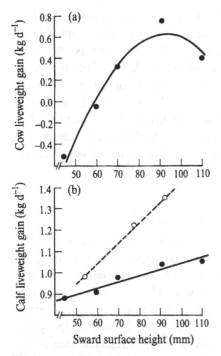

Figure 6.4. Effects of sward surface height on (a) cow and (b) calf LWG during May–August (●) and August–September (○) at Shotts, UK. (From Wright and Whyte, 1989.).

with 300 kg N ha^{-1} season^{-1} at Hillsborough, Northern Ireland (Chestnutt, 1992) showed a quadratic relationship with increasing SSH from 30 to 90 mm. Management decisions about SSH depend on whether young lambs are to be marketed or whether store lambs are to be produced. Mean SR to weaning decreased from 25 to 13 ewes ha^{-1} as SSH increased from 30 to 90 mm, and lamb LWG to weaning decreased from 808 to 584 kg ha^{-1} respectively. In this study a target LW of 40 kg (35 kg LWG) at 120 days would require 290 g LWG lamb^{-1} d^{-1} which would not be attainable under close grazing at this site.

Rather lower SSH appears possible at other sites. At Hurley, UK Penning *et al.* (1991) recommended the optimum SSH for continuously stocked swards of between 30 and 60 mm for ewes and lambs and no advantage in individual LWG was evident at SSH in excess of 60 mm. At the same site Orr *et al.* (1990) found that SSH close to 60 mm was desirable for both grass + N and grass–clover swards; the latter had an animal output *c.* 80% of the former. At Lanarkshire, UK Maxwell *et al.* (1994) found acceptable flock performance if *L. perenne* dominant swards were controlled to a SSH of between 35 and 50 mm during spring and summer, provided supplementation was offered if SSH fell below 35 mm. At Palmerston North, New Zealand Parker & McCutcheon (1992) recommended SSH of 50–70 mm during ewe lactation, and noted the higher intake of herbage by ewes rearing twins than by ewes rearing single lambs; wool growth was independent of SSH treatment. Ewes in good condition could achieve a satisfactory level of OM intake over the final month of pregnancy on closely grazed swards of 20 mm SSH, although ewe LWG increased with increasing SSH (Morris *et al.* 1993).

6.4.2 *Forage conservation and sward surface height*

The integration of grazing with setting aside pasture areas for conservation makes adjustment to SSH feasible, since this may be achieved simply by varying this buffer area, whose actual size will be partly dependent upon the seasonal growing conditions experienced. This may still be anchored within a basic system which ensures adequate provision of conserved forage for the winter. For example, half the pasture area may be reserved for spring silage and the aftermath grazed together with the unreserved area nominally grazed throughout, but a buffer area of varying dimensions which is conserved for silage on every occasion when uninterrupted growth of *c.* five weeks occurs may also be incorporated. Additionally, animals may receive supplementary feeding during the grazing season if adverse seasonal conditions or managerial misjudgement reduce SSH below an acceptable level.

Illius, Lowman & Hunter (1987) tested buffer areas varying in extent from 25 to 48% of the pasture, which were grazed by young cattle in Midlothian, UK to SSH target heights of 50 or 30 mm. The best compromise over four years was a SSH of 50 mm and a buffer area of 25–30% of the pasture area, which permitted substantial silage production from the buffer area and pragmatic grazing management of a high fertilizer N system. SR and N input may be varied and a common SSH maintained to give satisfactory animal LWG with differing outputs of silage conserved.

At State College, Pennsylvania Fales *et al.* (1993) used half the area of mixed pasture (*Dactylis glomerata, Poa pratensis* and *Bromus inermis*) for spring silage production and varied SR to 2.5, 3.2, and 3.9 cows ha^{-1}, supplementing these animals with grass silage if forage allowance was deficient. This led to equal outputs of milk cow^{-1} across SRs, despite the range in SR applied, but the surplus silage not fed during the grazing season varied from 8.1 to 4.7 t DM ha^{-1} as SR increased. When the effects on body condition are also taken into account the manager is able to align a suitable SSH with the economic returns from concentrate feeding. SSH may be employed as a management tool which mitigates risk in circumstances where varying inputs may be used to optimize farm return whilst maintaining the vigour and composition of the sward.

6.4.3 Herbage intake and sward characteristics

The factors controlling herbage intake are fundamental to the understanding of animal responses to variation in SSH (Gordon & Lascano, 1993). T.H. Stobbs (1973a, b) was one of the main pioneers of the concept that ingestive behaviour is modified by the structural characteristics of the sward. Animal intake may be viewed simplistically as the product of OM per bite, rate of ingestive bites, and duration of grazing. On the lax, tall, rather stemmy subtropical swards (*Setaria sphacelata* var. *sericea, Chloris gayana, Macroptilium atropurpureum*) with which Stobbs worked, sward bulk density and leafiness appeared to impose limitations to individual bite weight, and consideration of the rate of prehension biting and the daily duration of grazing led to his suggestion that a bite size of less than 0.3 g OM limited the herbage intake and milk output of his cows, which were loth to take more than 36 000 bites per day. On temperate swards of higher bulk density and leafiness Hodgson (1981) also emphasized bite size as the most significant variable affecting intake, but at that stage he discarded sward density and leafiness in favour of SSH as the predominant correlative factor which controlled the depth of biting (Barthram & Grant, 1984).

A field study (Penning *et al.*, 1991) at Hurley, UK in which *L. perenne* swards were continuously stocked with ewes and lambs to maintain SSH of 30, 60, 90 and 120 mm illustrates some aspect of this question. During the spring average herbage intake (Figure 6.5a) was substantially reduced in the 30 mm SSH treatment, especially at the beginning of the growing season when forage availability was low. Mean bite mass was estimated to increase from 45 to 155 mg OM bite^{-1} as SSH increased from 30 to 120 mm. However, compensatory grazing behaviour was evident. The duration of grazing increased in the low swards at the expense of time spent ruminating (Figure 6.5b). The rate of total jaw movements was remarkably constant across treatments (Figure 6.5c), but ewes on the SSH 30 mm treatment had a higher rate of prehension biting and a reduced rate of mastication; as bite mass decreased a lesser proportion of jaw movements were masticating movements. However, the rate of chewing and the number of chews per bolus were independent of treatment; the number of boluses per day were

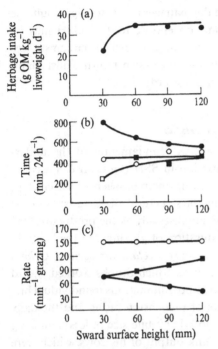

Figure 6.5. Sward surface height and (a) herbage intake of ewes; (b) time spent grazing (●), ruminating (■) and idling (○), and (c) rate of total jaw movements (○), prehension biting (●) and masticating (■) at Hurley, UK. (From Penning *et al.*, 1991.)

fewer in the severely grazed treatment, and the interval between swallowing each bolus and regurgitating the next bolus was larger. The reduced intake of ewes in the 30 mm treatment was reflected in poorer individual LWG of lambs and reduced ewe condition score; it is not usually feasible for sheep or cattle to compensate wholly for the effects of acutely low herbage availability by increasing duration of grazing and rate of biting.

Sward density may be involved as well as sward height in determining intake, as illustrated by a contrasting study (Burlison, Hodgson & Illius, 1991). Bite weight may be viewed as the product of the bulk density of herbage in the grazed stratum of the pasture and the bite volume, which in turn is the product of the depth of biting and the bite area. The combined effects of SSH and the bulk density of the grazed stratum are shown in Figure 6.6 for 17 swards of oats and perennial grasses of widely differing structure. The depth of grazing may be determined by the SSH above the level of dense pseudostems; in Kentucky, USA, cows stopped grazing

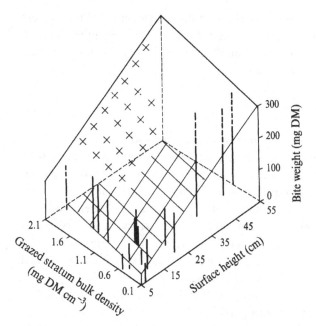

Figure 6.6. The combined effect of sward surface height and grazed stratum bulk density on bite weight of ewes grazing various swards. (From Burlison *et al.*, 1991.)

Festuca arundinacea pastures at 100–120 mm above soil surface, defined approximately by the tops of the pseudostems (Dougherty *et al.*, 1994). Wade *et al.* (1989) noted in France a progressive reduction in depth of grazing of *L. perenne* swards and a concomitant reduction in area per bite as cows reduced SSH.

Animal performance results from the product of herbage ingestion, the nutrient concentration of the herbage ingested, and the efficiency with which these are converted to animal product. The herbage on offer at very low SSH may be of low digestibility, but lax swards of high SSH may contain much senescent or stem material, as at Fermoy, Ireland (Stakelum & Dillon, 1989) and at Nishinasuno, Japan (Togamura, Ochiai & Shioya, 1993). This issue is sensitive to local pasture type and conditions; the Hurley study quoted above (Penning *et al.*, 1991) found similar levels of OM digestibility in continuously grazed swards maintained at differing SSH, whilst Hodgson's (1981) calves in Midlothian, UK ingested *L. perenne* material of much higher OM digestibility at a grazing height of 80 mm or above. The opportunity for selective grazing clearly influences the outcome of this question. The emergence of local studies of SSH responses on temperate pastures provides a powerful management tool for graziers, who are then able to target the desired total and individual animal outputs through manipulation of SSH, animal numbers, fertilizer inputs, purchase of supplements and fodder conservation policy.

6.5 Stocking method

The most desirable system of grazing management which has evolved over recent decades is continuous stocking with some variation in seasonal SR which brings forage supply more into synchrony with animal needs, and influences pasture vigour and botanical composition (Humphreys, 1991).

This affirmation is at variance with that of the proponents of more complex systems of grazing. These have a long history; in 1777 Anderson (cited by Voisin, 1959) advocated rapid rotational grazing over a 20-paddock system. 'Rotational grazing' may be used as a generic term which incorporates many stocking methods; the basic feature is that a group of animals grazes over more than one paddock and is moved from paddock to paddock in accordance with particular precepts.

The flexibility of the system increases with the number of paddocks, which is usually 3 to 15, but may extend to 50 in 'cell' grazing. The key questions for the manager are the duration and timing of grazing in particular paddocks, and the duration and season of rest interval between grazing of particular paddocks. The movement out of paddocks may be rapid, with a grazing duration of one day or less, as in strip (or 'crash', short duration or ration) grazing systems where the grazing area may be adjusted with a movable electric fence, preferably with a back wire to prevent animals grazing the recently grazed area. The rest interval may be a few days, usually 15–60, or may be as long as a year if plant regeneration is sought, as in 'deferred grazing', or if fuel is being accumulated for a fire. The movement of stock may be on a rigid calendar basis, or may be adjusted according to immediate growing conditions as in 'time control grazing' (TCG).

6.5.1 Objectives of stocking method

The declared objectives of different grazing systems are varied and sometimes contradictory but are embraced by the following:

Vigour and persistence of the sward

André Voisin (1959) was an influential exponent of multi-paddock grazing systems, and manufacturers and suppliers of fencing materials in many countries are greatly indebted to him. Three of his key ideas directed to the promotion of plant vigour and persistence were that (1) duration of grazing in a paddock should be sufficiently short to preclude animals grazing the regrowth of recently defoliated plants, which would deplete carbohydrate 'reserves' in the crown and roots, (2) duration of rest should be sufficiently long for carbohydrate reserves to be replenished and also (3) for the pasture to enter its 'blaze of growth' phase. Following severe defoliation there is a lag phase in growth before LAI has increased sufficiently to utilize incident radiation to bring about rapid carbon fixation as discussed in chapter 4 (for example, Brougham, 1956; Ludlow & Charles-Edwards, 1980). A long rest may enable soil seed reserves to be replenished for future plant replacement or incursion.

Botanical composition

Grazing may be directed to the control of unpalatable species or weeds, the regeneration of the perennial grass components at the expense of annual species, or to the maintenance of a legume component in mixed pastures. J.P.H. Acocks (1953, 1966) in South Africa advocated veld reclamation through 'non-selective' grazing; in this system duration of grazing was short and heavy to enforce consumption of the less acceptable species and discontinuous to permit subsequent recovery of the palatable species.

Increased utilization

A multi-paddock system is claimed to enable grazing to '. . . be timed to utilize the maximum inflexion part of the growth curve of forage, (so that) . . . higher stocking rates are possible from the outset' (Savory, 1978). Allan Savory (1978) suggested that short-duration grazing might safely double SR 'as long as adequate time control is brought into the grazing handling' and that a three-fold increase in SR is feasible in some circumstances. Savory's concept of 'holistic resource management' embraces management concepts additional to time control grazing, but the latter is central to his thesis of better herbage utilization. The subdivision of pastures implicit in a multi-paddock system (or alternatively in set stocking systems) leads to a more even spatial pattern of grazing.

Zero grazing, or cut-and-remove systems of forage feeding have shown in intensive systems a gain of *c.* 5–8% over grazing (Lowe, 1970), which does not cover the additional costs or compensate for the environmental damage associated with such systems, which retain popularity in many tropical countries for reasons of farm convenience.

Specific animal needs

A basic concept enunciated by many advocates is that rotational grazing enables the manager to present to the animal herbage of high nutritive value (Lowe, 1970). More specifically, paddock management may be directed to accumulating surpluses to meet particular physiological needs such as birth, early lactation or mating, or to anticipate a seasonal shortfall in feed supply, as in the New Zealand system of 'block grazing' which is directed to meeting animal requirements for winter grazing, achieved by reducing forage allowance during late lactation. This may also be achieved by seasonal adjustment of SR under continuous grazing.

'Creep' grazing ensures young animals have access to additional pasture, or to first use of pasture. This and systems with a long rest interval may aid the control of internal parasites (Lambert & Guerin, 1990), whilst long rotation grazing, involving the complete mustering of the host cattle, reduces the occurrence of the ectoparasite *Boophilus microplus* (Wilkinson, 1964). The incidence of facial eczema (*Pithomyces chartarum*) in sheep is reduced in lax pastures (Campbell, 1970).

Herd effect

Savory (1983, 1978) claims that herd trampling and herd 'excitement' are necessary to break soil capping and cryptograms in order to promote infiltration. Environments subject to long dry periods are termed 'brittle' and hoof action of the herd achieves the physical breakdown of litter, aiding nutrient cycling, and achieving seed burial. The concentration of animals in larger herds and herd movement are accomplished through a multi-paddock system.

Farm convenience

Rotational grazing simplifies many farm procedures, such as the reservation of paddock areas for fodder conservation, avoidance of stock trampling after irrigation, the segregation of animal classes, and health routines such as dipping and inoculation may be combined with movement between paddocks, which is facilitated as stock become accustomed to these movements and are handled more easily. The inevitable regular contact of the operator with the herd promotes pastoral care, and feed budgeting is simplified on small paddocks.

6.5.2 Evaluation of achievement of objectives

To what extent have the objectives enunciated above been achieved in farm practice?

Vigour and persistence

The focus of grazing management on the maintenance of carbohydrate reserves was discussed in chapter 4 and largely discounted. The concern that animals might graze recent regrowth has not been validated in field studies where normal SR is applied. In Natal, South Africa Barnes (1992) noted that during a 14-day grazing period in a three paddock system only 9% of the tillers of grasses of interest were regrazed during the period of stay. At Matopos, Zimbabwe the frequency of defoliation of marked tillers was similar under continuous and rotational grazing where stock had durations of 6 or 12 days in particular paddocks (Gammon & Roberts, 1978). Frequency and height of defoliation was also similar in four and eight paddock systems (Gammon & Twiddy, 1990).

It is not easy to make valid comparisons of the growth rates of pastures continuously or intermittently grazed; there are many reports of slightly greater herbage accumulation under rotational grazing (for example, Jones & Jones, 1989) but changes in sward structure under continuous grazing, leading to high tiller density and reduced rates of senescence, promote a homeostasis in which herbage yield is similar under a range of stocking methods (Ernst, Le Du & Carlier, 1980).

Botanical composition

Species composition has been successfully manipulated following critical studies of plant behaviour and demography. The seasonal life cycle of preferred and non-preferred species might be studied and grazing pressure adjusted to favour the seed production and regeneration of the preferred species (Lodge & Whalley, 1985). The legume component of the pasture might also be favoured by seasonal variation in grazing pressure (Jones, 1984, 1988; Fujita & Humphreys, 1992).

In New South Wales *Aristida ramosa* has been decreased and *Danthonia linkii* favoured by heavy summer–autumn grazing with sheep, which reduced seed production of the former, and winter–spring rest which enhanced seed production of *D. linkii* (Lodge & Whalley, 1985). In Queensland long-term exclosure and spring burning has restored the dominance of *Heteropogon contortus* over *Aristida* spp. (Orr, McKeon & Day, 1991). Resting of pastures to develop fuel for a hot fire has in many instances reduced the incidence of fire-susceptible woody species (Barnes, 1965; Hamilton & Seifres, 1982).

On the other hand, non-selective grazing has failed badly. 'Forced grazing of unpalatable plants usually results in extreme defoliation of the palatable species' (Gammon, 1978) and paired comparisons in Zimbabwe of properties with intensive short-duration grazing management (14–48 paddocks per herd) and of properties with extensive management systems showed no evidence of marked or consistent superiority of the former with respect to species composition or basal cover (Gammon, 1984). Table 6.3 summarizes southern African experience and indicates no inherent weakness of continuous stocking in the maintenance of desirable botanical composition. In tall grass prairie in Oklahoma various regimes of short duration grazing did not affect botanical composition (Gillen *et al.*, 1991), whilst in Texas midgrass cover was favoured by moderate continuous stocking (Taylor, Garza & Brooks, 1993). In intensively managed pastures the lower tiller density and open swards associated with rotationally grazed pastures may encourage the ingress of less favoured species (Ernst *et al.*, 1980).

Much effort is directed to the maintenance of the legume component of pastures. Some selectively eaten plants, such as *Medicago sativa*, require intermittent grazing for their survival (Moore, Barrie & Kipps, 1946), and examples were quoted earlier of seasonal SR adjustments leading to successful legume adaptation. The pattern of seasonal selectivity requires study; many tropical legumes are positively rejected during the wet season (Stobbs, 1977; Böhnert, Lascano & Weniger, 1985) and the apparent preference of animals for *T. repens* (Curll *et al.*, 1985) is modified by previous dietary experience (Newman, Parsons & Harvey, 1992).

Increased utilization

Increased herbage accumulation associated with high LAI arising from a long rest interval is only achieved through lower levels of herbage ingestion; further, at the upper levels of underutilization herbage accumulation is decreased, as discussed in chapter 4 (Parsons, Johnson & Harvey, 1988). The proposal that time control grazing can significantly increase SR has no basis in science, and exponents of this claim greatly threaten the conservation of grassland resources. Grassland scientists who have tested multi-paddock grazing have been unable to support Savory's claims that SR may be safely doubled (Heitschmidt *et al.*, 1982; Pieper & Heitschmidt, 1988; Bryant *et al.*, 1989; Barnes, 1992; O'Reagain & Turner, 1992).

Specific animal needs

Grazing management directed to particular physiological needs is a valid emphasis. The caveats are (i) aged feed is of lower nutritive value than young herbage, and this accounts for part of the reduced LWG often recorded in multi-paddock systems (Bryant *et al.*, 1989; Volesky, Lewis & Butterfield, 1990); (ii) selective grazing is needed to improve animal diet, especially where C_4 grasses are grazed (Böhnert, Lascano & Weniger, 1986). Time control grazing may allow selective grazing if the SR is sufficiently low and the rotation sufficiently rapid to resemble continuous stocking. It is salutary to discover that the advantage gained by control of ectoparasites through rotational grazing can be entirely offset by the lower nutritive value of the herbage offered under this system (Johnson & Leatch, 1975). On the other hand property subdivision *per se* (which is not necessarily linked to any particular stocking method), reduces energy expenditure on walking and may benefit LWG (Hart *et al.*, 1991).

Herd effect

Scientific evidence is entirely contrary to the claims made above; herd trampling causes soil compaction, reduces infiltration, and greatly increases runoff and soil erosion (Warren *et al.*, 1986; Phular, Knight & Heitschmidt, 1986; Welz, Wood & Parker, 1989). Burial of small-seeded grasses by hoof cultivation may be too deep to be effective in facilitating seedling regeneration (Hacker, 1993); establishment of *Panicum coloratum* under continuous grazing was greater than or equal to that under short duration grazing (Bryant *et al.*, 1989).

Farm convenience

The earlier affirmations in section 6.5.1 about farm convenience are valid.

6.5.3 Summation

Ernst *et al.* (1980) reviewed grazing studies in western Europe and concluded that the benefit of rotational grazing averaged *c.* 1.5% for dairy production and *c.* 6% for beef production; these increases would not pay for the added fencing and management costs attributable to rotational grazing. Since then the development of controlled continuous grazing systems using SSH have given graziers pragmatic management systems which have been widely accepted.

For tropical and subtropical pastures an analysis of 60 grazing

Table 6.2 *Comparison of effects of stocking method on individual performance of animals grazing tropical and subtropical pastures (% experiments)*

Proposition	Yes	No	No difference
Continuous grazing increases animal performance relative to rotational grazing	51	17	32
Long rest interval is detrimental to animal performance relative to short rest interval	33	17	50
Long duration of grazing in the rotation cycle is detrimental to animal performance relative to short duration of grazing	45	9	45
Long rotation cycle is detrimental to animal performance relative to short rotational cycle	14	0	86

Source: Humphreys (1991).

experiments for sheep, beef and dairy cattle (Table 6.2) indicates the inferiority of rotational grazing relative to continuous grazing, and a high proportion of studies in which stocking method has no consistent effect on animal production.

A review of 50 grazing studies in southern Africa (O'Reagain & Turner, 1992), some of which are incorporated along with studies from other continents in Table 6.2, also suggests no advantage either for generic rotational grazing or for multi-paddock systems for both animal production and range condition (Table 6.3). Skovlin (1987) has reviewed the failure of graziers in private farming and communal areas in Zimbabwe to adopt short duration grazing, despite serious extension efforts. Hoffman (1988) in discussing the options for management of the semi-arid Karoo region suggests that the institutional proposals have ignored the unpredictable nature of this environment and have propounded untested or untestable theoretical frameworks; it is surprising that as recently as 1985 some southern African scientists (for example Tainton, 1985) implied that SR is of secondary importance to grazing method. Similarly, the advocacy of time control grazing has recently emerged in Australia, and McCosker (1994) dismisses scientific rigour in favour of selective anecdotal evidence. He discounts the results of grazing experimentation because of the claimed lack of flexibility in the timing of stock movement, although this in fact was well incorporated in some multi-paddock grazing studies, such as those of Bryant *et al.* (1989) and Volesky *et al.* (1990). It should be recognized that many US range scientists in the 1970s and 1980s embraced short-duration grazing

Table 6.3 *Comparison of effects of stocking method on animal production and range condition in southern Africa (% experiments)*

Proposition	Yes	No	No difference
Continuous grazing increases animal production relative to rotational grazing	39	30	30
Continuous grazing improves range condition relative to rotational grazing	14	23	64
Multi-paddock rotational grazing improves animal production relative to rotational grazing over few paddocks	14	29	57
Multi-paddock rotational grazing improves range condition relative to rotational grazing over few paddocks	0	33	67

Source: O'Reagain & Turner (1992)

with enthusiasm as a potentially valuable concept, and have only discarded it after serious and long-term study.

There are immense differences in animal productivity and in sward composition arising from variation in SR, forage allowance and SSH. When these are contrasted with the small and inconsistent differences in animal productivity and resource conservation arising from variation in stocking method, some scientists should be grateful that graziers in the main rejected their advice about stocking method. It is of interest to speculate why grassland science took wrong turnings for so many decades. D.L. Barnes (1992) states in his review: '. . . past advice for the management of sour veld was seriously flawed . . . the factual base for sound veld management is as yet incomplete . . . The present appraisal shows that multi-paddock grazing procedures can be regarded as a misuse of capital and an unwarranted expense'. O'Reagain & Turner (1992) suggest in a crushing indictment of grassland work in southern Africa that 'basic questions in rangeland management remain unanswered. This could be attributed to the unfamiliarity of many workers with the basic methods and philosophy of science'.

This chapter is incomplete in that discussion of the kinds of animals grazed, another aspect of stocking method, has been omitted. Mixed grazing of domestic animals, especially directed to minimizing sheep damage to pasture resources and to complementary use of mixed pasture (Lambert & Guerin, 1990) has gained momentum. Concurrently, there is increasing interest in the use of grassland resources by large grazing ungulates and the conservation of animal biodiversity (Barnes, 1993; Duncan & Jarman, 1993), as mentioned in chapter 1.

The focus of this chapter is on the pasture–animal interface and the development of a sustained synchrony between the needs of the animal and of the pasture for survival and production. This requires further development through the application of various approaches (Humphreys, 1991) to overcoming discontinuity in forage supply.

7
Innovation, optimization and the realization of change

7.1 Introduction

Holistic science has always featured in grassland improvement, and in the last three decades there has been a major change of emphasis towards the systematic modelling of the component processes of grassland production and their interaction with each other, and towards systemic studies which would improve the prospects for the realization of change.

At the IV International Grassland Congress R.G. Stapledon (1937) stated: 'Grass (and when I say 'grass' I mean, of course, grass and clover) properly used ensures soil fertility, grass marries the soil to the animal and the solid foundation of agriculture is the marriage of animal and soil'. Although the substance of research papers at this time was focused on reductionist science and the innovations which might emerge, most grassland agronomists were generalists with close links to the farming community. The agenda of the International Grassland Congress dealt with aspects of the soil–plant–animal system and there was little space for the social sciences, but significant numbers of farmers attended the meetings and the transfer of technology was a central goal. P.V. Cardon (1952) encapsulated an ethos: 'These measures . . . seek only to subjugate by exact knowledge the natural obstacles that impede human progress towards a more abundant life'.

R.O. Whyte (1960) enunciated a wider concept than was evident amongst many of his contemporaries in suggesting that in grassland development 'many of the limiting factors will be brought into relief – the natural ecological factors, the factors related to social and economic conditions, and the factors related to fundamental ways of life, tradition and religion'. At this VIII International Grassland Congress W.J. Thomas (1960) pointed to deficient '. . . approaches as currently practised by agricultural scientists and agricultural economists [which do not] give us completely adequate answers to farmers' problems . . . . The grassland scientists . . . can provide answers as to the effect of many specific actions . . . without being able to answer what will be the overall effect of adopting several actions at the same time, which is essentially the farmer's problem'.

The word 'systems' first appears in the title of a paper at an International Grassland Congress in 1966 (Baker, 1966). Jones & Baker (1966) described the functions and methods of integrative research as the construction of both simple physical and complex mathematical models of grassland systems, the testing of these systems and the definition of areas in which 'conventional analytical research' is needed. They instanced a physical model of intensive beef production arising from the adoption of various research outputs, and which was associated with the formation of a Systems Synthesis Department at the Grassland Research Institute at Hurley, UK (Brockington, 1974).

There were three plenary papers about systems at the XI International Grassland Congress. R.W. Brougham (1970) argued for studies of the whole soil–plant–animal system in an ecological context of field experimentation. C.R. Spedding (1970) was reassuring about the ease with which the complexity of grassland systems might be handled; most natural systems were well buffered to reduce the consequences of variations. Scientists needed to identify the relatively few important factors which determined the efficiency of production, and 'the apparently greater complexity of whole systems is largely an artefact, due to the retention of inappropriate component units'. He saw the function of models as that of organizing data into usable knowledge, of improving understanding, of assisting scientists to identify gaps in both knowledge and understanding and of permitting prediction of the effects of change in component processes.

George M. Van Dyne (1970) was a dominant evangelist at this Congress, and with his school at Colorado State University, exerted a powerful influence in promoting research on grassland systems until his untimely death in 1981. Perhaps his central thesis was that: 'Detailed ecosystem studies are needed before existing theories can be applied fully to optimize biological productivity for man'. He expanded the concepts enunciated so far into his systems approach which included: (i) compiling, condensing and synthesizing much information concerning the system components; (ii) detailed examination of the system structure; (iii) translating knowledge of components, function and structure into models; and (iv) using models to derive new insights about management and utilization. Van Dyne gave illustrations of compartment models in which the mathematical relationships of energy flow were expressed, and listed conventions for a macro-model of grassland incorporating vectors of ecosystem components, of input functions, and of properties together with sets of control functions and of processes. He emphasized the requirement for successive revisions to achieve a realistic simulation which might be extended as an instrument for manipulating the system, forecasting changes under natural and artificial stresses and improving the planning of long-term field research through sensitivity analysis.

The development of 'hard' systems derived mainly from the thinking of engineers who provided a quantitative, mathematical and reductionist basis for their operation. The marriage of these hard systems approaches with the people component of the grassland ecosystem had to await more recent Congresses.

7.2 Innovation and optimization

7.2.1 The role of reductionist science and of modelling

The major advances in grassland improvement have arisen from the application of reductionist science to the understanding of individual components of the grassland ecosystem. The increased outputs which are memorable arise from particular innovations: the introduction and identification of elite legume germplasm, such as *Stylosanthes hamata* in Queensland and Thailand; understanding the photosynthetic pathways of C_4 and C_3 grasses; the knowledge that soil nitrate inhibits rhizobial N fixation; the involvement of the micronutrient Mo in N fixation and the effect of soil pH on its availability; the greater reliability of legume sowings once hardseededness is manipulated; the photoperiodic control of seed production; the action of rain drops eroding bare soil; the controlling influence of pH on the quality of silage made from herbage. These disparate examples might be contrasted with the lesser gains which have arisen from the optimization of inputs controlling processes of plant production, originally derived from engineering and industrial land systems (Brockington, 1974). The financial gain from applying 200 units of N to a sward, rather than 250 units, may be real but the response surface may be rather flat.

Systems thinking is a way of interpreting 'reality' in which any individual component is viewed as being part of a larger whole (Breymeyer & Van Dyne, 1980; Jones, 1988). The boundaries of the system may be drawn at any level, and usually encompass the processes which impact directly on the things of particular interest, or under control in the particular study

(Checkland, 1981). The effect of reducing the pH of silage, or of a new additive in increasing its acceptance and digestibility and reducing nutrient losses, may be measured in a feeding experiment and reported together with appropriate blowing of trumpets. However, in assessing the impact of this innovation on the net outcome from the farm system the observer also needs to know the preceding effect of withdrawing an area of the farm for silage making on the performance of the animals grazing; further, the subsequent behaviour of unsupplemented animals would need to be compared with that of the animals previously receiving the superior silage, since compensatory weight gain may completely erode their earlier advantage. This example is given to indicate one reason why the outputs of reductionist science have often been incompletely applied, and this has been a factor in promoting the growing interest of scientists in systems thinking.

A more fundamental imperative intrinsic to the activity of science is to build a model which places the process under study within a picture of reality which indicates the strengths and the nature of the pathways which control its operations and also its connexion with other processes. Early grassland modellers, such as W.G. Duncan (1966), were driven by intellectual curiosity and the desire to build elegant models which depicted the mathematical dimensions of processes such as photosynthesis.

Modelling, according to Wilson & Morren (1990), is used to communicate complex interrelations and concepts about the meaning of something, as a novel construct for the search for new insights into the behaviour of a system or as a simulation for the evaluation of alternative strategies. Grassland scientists have found modelling an aid to reasoning and a useful method of identifying the gaps in current knowledge which might be addressed in future research. Models may be classified as iconic (or physical), which attempt to mimic simple or complex reality, or as symbolic (or conceptual), which model ideas of relationships or properties which control functioning. Mathematical models may be deterministic, characterized by assumed certainty and predictability (as illustrated in chapter 4 by Parsons, Johnson & Harvey, 1988), or stochastic, which deal with probability (as illustrated in chapter 6 by Scanlan *et al.*, 1994), and may also involve queuing, inventory, simulation and gaming.

Seligman (1993) realistically suggests that a biological simulation model 'cannot predict the future, replace experiments designed to discover biochemical pathways, ecological processes and site-specific responses to manipulation, [or] replace subjective assessments and value judgements that enter into many critical management decisions'. On the other hand, it may 'enable preliminary assessment of the effects of new conditions or new techniques on system response, [and] provide a means to explore systems behaviour and so help to distinguish between sensitive and stable properties of biological systems'. He states further that: 'it is on the opportunity to explore the response space of a system and discover distinctive behavioural patterns at the appropriate organizational level that the appeal and the promise of most agricultural modelling rests'.

A recurring theme in recent literature is the necessity to distinguish models which are directed to enhancing the research worker's depiction of reality from models whose purpose is to assist the producer in the management of grassland to achieve desired outcomes. There are many examples available of biophysical models which work in the farm situation and which are enhanced by the capacity to simulate the output from, say, 100 individual rainfall years. For instance, a key area for producers is the alternative pathways to successful marketing of sale animals of suitable weight and carcass condition at an agreed date set by meat processors. Pleasants, McCall & Sheath (1995) assessed the uncertainty and variability associated with particular feeding options so that farmers may manage better the intrinsic and differing variability in feed supply on their farms to achieve their desired objective.

Farm management modelling directed to a combination of biophysical and economic outcomes has become a major industry. Two recent examples are mentioned here. Nicholson *et al.* (1994) developed an optimization model for dual (milk–beef) production in the humid lowlands of Venezuela which was based on deterministic, multi-period linear programming. They calculated the maximized discounted herd net margin using coefficients for various feeds and animal nutrient requirements when varying animal, forage and purchased feed activities. This displayed the superior results from increasing the use of the locally available feeds, urea and molasses.

A combination of economic and environmental goals was employed by Fiske *et al.* (1994) in evaluating alternative grazing management and fertilizer strategies for beef cow/calf production in West Virginia. The assessments were made in terms of profitability, economic risk and soil loss. Weighted goal programming determined the outcome, which was sensitive to the weights given to each of the three factors; this would mimic the varying attitudes of producers and the emphasis given to environmental sustainability. Walker (1993) refers to attempts to develop models for grazing management derived from optimal control theory which would maximize a complex function of welfare based on long-term net economic benefit.

7.2.2 The development of decision support systems

Decision support systems (DSS) have evolved in response to the need to create 'mechanisms to develop and transfer technology that meets the needs of individual landholders and overarching values of society' (Stuth *et al.*, 1993 to whom this segment of the chapter is indebted). The great expansion of the knowledge base available to farmers and land managers has prompted new techniques for accessing this information (Whittaker, 1993), avoiding indigestion, and guiding heuristic (seeking knowledge for oneself) activity. The roots of these systems for grazing lands were in the financial analyses of agricultural economists and the simulation modelling of resource scientists. The wide nature of the databases involved and the diversity of players involved are shown in Figure 7.1. Organizations provide databases on the biophysical aspects of production and of the environment, and these supplement the information the landholders have about their own properties, their previous performance, and the nature of their resources. The significance of the landholders' activity in building their own databases, (for example using the computer package GRASS CHECK (Forge, 1994)) is emphasized. Policy analysts associated with governmental and public institutions provide input on communal concerns and on externalities which distort the producer's market and influence decision making. The technical specialist may work alongside the private consultant or extension agent in developing models from the databases for a long list of potential applications to aspects of production and its sustainability, and the closeness of the links with the landholder in defining the problems and determining the framework for their solution, amelioration or optimization is crucial to the success of DSS.

The central issues (Stuth *et al.*, 1993) are (1) the adequacy or applicability of the technology embraced by the DSS; (2) design considerations of the DSS; and (3) the requirements for user adoption and implementation. The knowledge base derives from data characterized as remote access, distributed and local site-specific, together with data transfer utilities. A database manager may be needed together with a manager dealing with the development of strategic, tactical and operational applications, whilst the interface with the decision maker may certainly require additional specialists. Functional knowledge of developers (of DSS) must go beyond understanding natural processes, systems analysis programming, economics, mathematics and statistics to incorporate human sciences, such as organizational behaviour and cognitive learning. The experience of Stuth *et al.* (1993) is that the main users of DSS currently are technical advisers

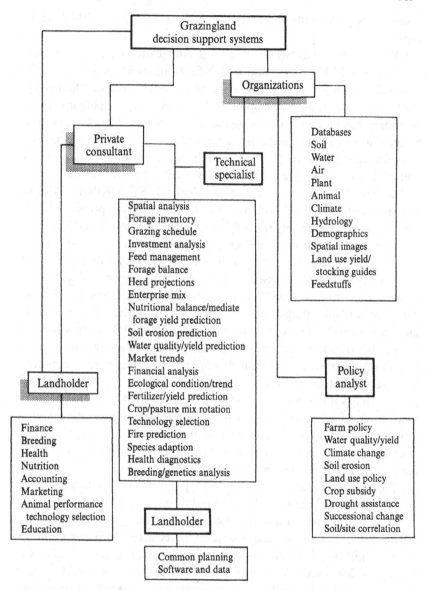

Figure 7.1. Categories of users and general listing of associated applications often found in decision support systems (DSS). (From Stuth *et al.*, 1993.)

working with landholders, and using soft DSS approaches until a problem arises in which decision making is facilitated by the application of a hard systems process. The primary step is to determine whether a computerized DSS is really needed and is feasible, and this decision will be modified by assessments of the skills, characteristics and declared needs of the potential users.

DSS is illustrated by the Grazing Land Application packages developed at Texas A & M University and USDA Soil Conservation Service (Figure 7.2). This DSS has animal, plant, soil, land use stocking guide, plant growth and transect data bases. The applications focus on developing inventories of grazing land forage, calculating the impact of mixed stocking, defining herds, determining the forage balance, scheduling grazing, managing feeds, analysing and mediating the nutritional balance, assessing managerial risk and analysing the long-term land improvement investment value.

Another illustration is the New Zealand STOCKPOL (McCall, Marshall & Johns, 1991), which assists the producer to optimize SR relative to pasture growth, and to establish feeding programmes, fertilizer practice, forage conservation and cropping. In Australia RANGEPACK (Stafford Smith & Foran, 1990) is directed to property-level management and to regional policy assessment, embracing economic feasibility, a Geographical Information Systems (GIS) oriented database for paddock management, and probabilistic treatment of climate. The Queensland Department of Primary Industries (Clewett *et al.*, 1991; Ludwig, Clewett & Foran, 1993) has developed a family of packages dealing with savanna woodland management, stocking determinations, structuring of beef herds and the integration of forages. Embedded expert systems are a valuable adjunct to assist the producer to describe expected responses to particular treatments, to select an array of technologies which are pertinent or to weight the indices in a complex relationship.

Some research funding bodies have been concerned that the investment in developing DSS has been ineffective since these packages are used by few producers (Cox, 1996), even where a high proportion of the target producers (as in Texas and in Queensland) have their own personal computers. The value to professional research workers of developing models is unquestioned, but Cox (1996) suggests that there has been (to 1995) a fundamental category mistake of confusing DSS models as a guide to practical action with process models suitable for the advancement of research. Organizations may have been too focused on outputs (computer packages) rather than on outcomes (changed and more efficient farm practices) (Cox *et al.*, 1995).

Some of these difficulties may arise from basic differences in personality type between farmers and research workers. Q.R. Jones (personal communication, 1996) used the Myers–Briggs Type Indicator (Briggs Myers & McCauley, 1985) to contrast farmers (Queensland rural leaders, predominantly cattle producers) with research scientists working in the

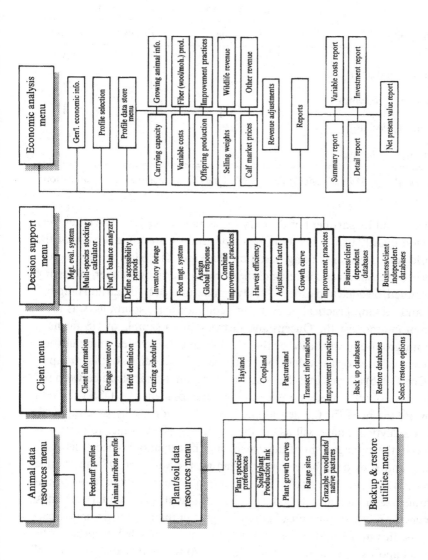

Figure 7.2. Components of the Grazing Land Application decision support systems developed at Texas A & M University and USDA Soil Conservation Service. (From Stuth *et al.*, 1993.)

Table 7.1 *Distribution (%) of personality types (Myers—Briggs Type Indicator) among research workers and farmers.*

	Research workers	Farmers	
Personality type	Queensland	Queensland	USA
SJ	30	45	50
(Sensing, Judging)			
SP	3	21	26
(Sensing, Perceiving)			
NF	7	11	4
(Intuitive, Feeling)			
NT	60	23	20
(Intuitive, Thinking)			

Source: Briggs Myers & McCauley (1985); Jones, Q.R. (personal communication, 1996).

Queensland Department of Primary Industries (Table 7.1). The latter were predominantly of the NT type, characterized by an intuitive view of the world and decision making based on thinking. The SJ type, viewing the world by sensation and relating to the world in a judging manner, preferring a planned, decided and orderly life style underpinned by detail and tangibility and seeking practical answers and commonsense solutions, was most common amongst the Queensland farmers and a sample of US farmers. Strong group cultures based on these temperaments impede communication when it is based on the values, perspectives and preferences of that group.

Cox (1996) affirms that the 'insertion of an analytical phase between the construction of a process model and the development of a DSS will often obviate the need for the development of a DSS for use by practitioners'. This analytical phase will assess the sensitivity of outcomes to perturbations in the management of the system and the extent of homeostasis operating and provide a basis for determining the value of DSS in routine decision making. The next issue is the use of an appropriate resolution in problem definition and the validation of simulated versus actual outcomes, where the noise in commercial practice is greater than in experimental treatments; this may lead to a mismatch between the intended purpose of the model and its resolution. Finally, Cox (1996) warns that the professional responsibility of scientists requires that craft skills are used sensibly and that outcomes are evaluated critically; argument by authority will not necessarily validate the belief that the outcome of a course of action is readily accessible and unproblematic.

Cox *et al.* (1995) define good models as 'accurate, simple and having explanatory power'; the balance of these attributes depends on whether the model is used 'for a technical, a practical or an emancipatory purpose'. Farming practice issues may be best handled by building on farmers' own heuristic models, whilst emancipatory issues need educational models which stimulate insight into system behaviour, as discussed in the following sections of this chapter.

7.3 Soft systems and the realization of change

7.3.1 The linear transfer of technology and farming systems research/extension

A long-lived model for creating change in farming practice is the linear transfer of technology (Jiggins, 1993). This depends for its success upon effective communication between research scientists and extension agents to whom research outputs are delivered; extension agents explain these to farmers and adapt them to local conditions of production, identifying the innovators who are most likely to be receptive to new ideas and technology. A 'trickle-down' effect to other farmers is expected. It is hoped that gifted scientists with a commitment to agriculture are likely to produce outputs relevant to the improvement of farm practice.

This model has been criticized on many grounds. Institutions may act as if they were closed systems with little attention paid to the quality and nature of the relationship between the system and its environment (Holt & Schoorl, 1990), although sufficient gestures may be made in this direction to protect the survival of a public-funded organization. In consequence, the activities of both research and extension workers may lack relevance to the real constraints to farm production and environmental protection and to the goals of landholders. The location of research may be heavily weighted towards the research station and laboratory rather than to on-farm sites. Issues of power and need in development are not addressed in this model (Scoones & Thompson, 1994), which has tended to transfer technology in an unbalanced way to the wealthier farmers.

The model may not be gender-neutral; Jiggins (1993) instances the development of legume fodder banks in West Africa with an inherent expectation that these would enhance the dry season grazing of pregnant and lactating cows which provided income from milk for women. In fact, the pastures were mainly used for late dry season grazing, especially for weak and sick animals, and then for the whole herd. This had the overall effect of increasing income for men from animal sales.

One response to the inadequacy of the linear transfer model has been the widespread development of Farming Systems Research (FSR) or Farming Systems Research/Extension (FSR/E). This has sought to expand the notion of an agricultural, biophysical system to the production and consumption issues faced by the farm family in its social context (Cox *et al.*, 1995), and attempts participatory technology development which emphasizes the joint design and evaluation of novel technological components by both researchers and farmers and the development in partnership of appropriate models: 'models become tools for managing within the communication process as we begin the delicate task of re-articulating two cultures that have grown apart'.

Farming Systems Research has been ineffective in many situations, but Biggs (1995) gives examples from developing countries where successful outcomes have arisen where 'the development and use of research methods were linked to the political, economic and institutional context in which they were developed and used'. A wider view of democratic participation in the learning process has been crucial. The orientation of Farming Systems Research to smallholders has been extended in an agroecological context to embrace the complex interactions between the components and processes of natural systems and human activities associated with agriculture (Oberle, 1994), and these involve the whole community as well as the farmer.

7.3.2 *Experiential learning*

Since ways of thinking guide action, the route to changed action is through the transformation of ways of thinking, which may arise from considering new metaphors or entering into dialogue (McClintock & Ison, 1994). The notion that communication is a physical process, as in the transfer of information electronically, has given place to the realization that people do not share a common experiential world, but may have conversations about overlapping worlds of experience as their different pictures of reality intersect (Russell & Ison, 1992). The iterative cycle of experiential learning (Kolb, 1984, applied by Bawden *et al.*, 1985) has become a key idea in the processes of agricultural research and community development. Arising from immediate concrete experience (Figure 7.3) the participant engages in reflective observations ('what is there?') which leads to 'a preliminary ordering of the elements, a tentative identification of objects, a proposed defini-

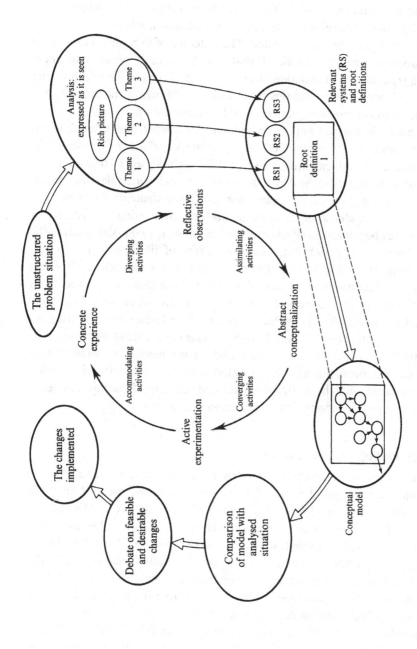

Figure 7.3. The relationship of the experiential learning cycle (inner) to soft systems methodology. (From Packham, Ison & Roberts, 1988.)

tion of relationships' (Wilson & Morren, 1990) accompanying the question: 'what does it mean?' Assimilating activities with convergent thinking ('what is important?') develops abstract conceptualization, which enables active experimentation to be formulated. The outcome of experimentation leads to accommodating activities ('what can be done?') which implement change in the actual situation. The lower half of the inner circle in Figure 7.3 deals with comprehending an abstract world, whilst the upper half is directed to apprehending a 'real' world (Wilson, 1988).

This cycle may be applied in reductionist science and takes different expressions in soft systems methodology (Checkland & Scholes, 1990). The unstructured problem situation (Figure 7.3) is analysed so that questions about such matters as the deployment of resources in particular operational processes, the planning procedures within existing structures in an environment where wider systems impinge, and the monitoring and control of these may be addressed (Packham, Ison & Roberts, 1988). This leads to the development of a rich but divergent picture of the situation with predominant themes identified. The next step requires that relevant systems and root definitions be devised. A root definition (Smythe & Checkland, 1976) may utilize the mnemonic CATWOE where the combined elements are: Customers, Actors in the situation, Transformations reflecting an accord on future operation, a *Weltanschauung* (world view) with open, participative decision making, Owners, and Environmental and wider system constraints. From this a conceptual model (with sub-systems) arises, which is compared with the analysed situation; discussion of feasible and desirable changes leads to the implementation of change.

7.3.3 *Participatory rural action*

A key activity in developing models for grassland improvement is the formulation of problems, which has overtaken the idea of problem identification (Ison, 1993). Rapid Rural Appraisal (RRA, McCracken, Pretty & Conway, 1988) has come to be regarded as a rather extractive process directed to benefiting researchers (Scoones & Thompson, 1994), and participatory rural action (PRA) which involves local people in the decisions about future action based on learning outcomes is counted more desirable. Pastoral development has suffered from the poor judgement of planners of agricultural development and planning for and not with producers has led to unfortunate consequences in sub-Saharan Africa (Waters-Bayer & Waters-Bayer, 1994). Value given to local people as actors rather than as targets for the transfer of technology leads to the greater empowerment of rural communities. This requires that local experience be elicited (Osty & Landais, 1993) and that rural people's knowledge be accorded equivalent value to scientific outputs (Ashby, 1990). Indigenous knowledge is manifold, discontinuous, dispersed and episodic (Scoones & Thompson, 1994), but constitutes the basis both for stakeholders formulating a consensus which conceptualizes a local problem and for the development of outcomes which lead to an improved situation.

These contemporary approaches are illustrated from a multi-disciplinary farming systems research/extension project in Santa Catarina, southern Brazil (Pinheiro, Pearson & Ison, 1994), where the current model is contrasted with an alternative participatory paradigm which is based on the constructivist theory of communication (Table 7.2). This seeks to build a richer picture of the context which is derived from the farmers' perspectives and participation, and to identify groups of farmers with enthusiasm for action. The systems understanding evolves from the interaction of farmers and experts; there is a focus on anthropological, behavioural and cultural issues and the farmers' view of potential improvements, whilst the experts are regarded as being within the system. Participative action evolves which emphasizes self-development of farmers, with experts acting as facilitators of the learning community of farmers.

Such a model requires attitudinal changes in scientists trained in more authoritarian modes. Russell & Ison (1992) and Cox *et al.* (1996) proposed an innovative metaphor: an invitation to the dance. They found that the rules farmers use in decision making did not fit the conventional decision tree structure, and suggest that the dance is a more appropriate metaphor for articulating change in which researchers and farmers communicate. Researchers engage in the dance to learn the rules; when new partners are engaged, rules may be proposed which were learnt from previous partners. As researchers become more skilled in the steps, they may influence the conduct of the farmers' own dances.

Externalities continually impinge on the decision making of farmers. Pastoralists in South Australia are more worried about the intervention of an ignorant environmental lobby than about the vicissitudes of economic conditions (Holmes & Day, 1995). In reviewing the World Bank's involvement in pastoral development in West Asia and North Africa de Haan (1994) states: 'subsidised crop cultivation (through high intervention prices, and subsidised tractors and fuel prices) caused key resource areas for the pastoral systems to be ploughed up, and subsidised animal feed prices interfered with the natural equilibrium, especially in times of drought'. Societal concerns are simultaneously directed to achieving economic

Table 7.2 *A current model of farming systems research in Santa Catarina, Brazil, and alternatives embracing a constructivist communication approach*

Current model	Alternatives
Diagnostic Search for objectivity/single reality (positivism) Data collection with focus on quantitative information Emphasis on problem identification and prescribed solutions (outsiders' interpretations) Focus on closed questionnaires followed by formal interviews (what people do) Predominance of 'rapid rural tourism'	*Context understanding* Built a 'richer picture' of the context. Focus on dialogue and exchange of qualitative information Emphasis on problem formulation and situation improvement (farmers' perspective) Participative methods such as participatory rural appraisal, semi-structured and focus-group interviewing (how people make sense of what they do)
Typology Identification of homogeneous groups (types) of farming systems Quantitave analysis based of farmer's physical and financial resource levels	*Group organization* Identification of groups of farmers with 'common enthusiasm for action' Qualitative analysis based on farmers' objectives, needs, attitudes, and theories
Monitoring Farming systems selected and monitored by experts (with farmers' agreement) Focus on physical, biological, economic, and financial aspects (what is happening) Search for what is wrong (experts' views) Data gathering and information giving viewed as distinct (transmission mentality) Experts remain outside the systems, deciding and controlling most of the actions. Farmers participate as observers and collaborators	*Systems understanding* Farming systems selected and studied in conjunction with farmers and experts Focus on anthropological, behavioral, and cultural aspects (why is it happening) Search for improvements (farmers' view) Farmers' experience and perspectives are valued, with information and decisions being shared (not only transferred) Experts reflect and interact within the systems, facilitating a network of conversations

Development

Central source of knowledge/power (top-down)

Participation is a project goal. Predominance of hierarchical and consultative participation ('power over' and 'power with')

Knowledge viewed as 'absolute truths'. Emphasis on conventional scientific research and extension (e.g., traditional trials, field days)

Research 'on' things and 'on' people

Learning, problem solving, and research viewed as distinct processes

Development is more dependent. Most actions and recommendations are based on expert's interpretations. Focus mainly on on-farm issues (production and productivity)

On-farm experimentations usually designed and controlled by experts. Farmers participate more as collaborators

Participative action

Knowledge/information are socially constructed and power sharing is encouraged

Participation is viewed as a process. Predominance of autonomous participation ('power to')

Tacit knowledge and ignorance are explored. Focus on informal and participative activities (e.g., story telling, community theatre)

Research 'with' people (co-learning process')

Research, technology, and development result from a cooperative process of inquiry

Emphasis on self-development. Farmers set their own agenda and priorities of action.

This may include issues 'beyond the farm gate' (e.g., marketing and political aspects)

Responsibilities for action are shared between farmers and experts. The latter act as facilitors of the learning community

Source: From Pinheiro *et al.* (1994).

growth, social equity and environmental protection (Nores & Vera, 1993), and the development of compatibility between these goals requires accommodating behaviour and skills in conflict resolution.

7.4 Conclusion

These contemporary approaches to the realization of change in farming practice offer great (but as yet unfulfilled) promise, and many observers are cautious in their expectations that the processes will prove any more effective than the discredited linear transfer of technology model. There is a similar scepticism that investment in the development of systems and of improved resource management is yielding greater benefits than reductionist research which results in innovations. Seligman (1993) takes the Darwinian view that the sophistication, relevance and utility of system models should improve with time, 'even if by selection pressure alone'.

The balance of effort in grassland improvement will be found by maintaining a spiral of interconnected learning cycles (Wilson & Morren, 1990) covering the spectrum from reductionist basic science, applied science and technology, and the application of hard systems through to the holistic study of soft systems (Figure 7.4). Each of these types of activity is only effective in grassland improvement as its interdependence with the others is recognized. A major advance from early International Grassland Congresses is the conceptualization of the left-hand side of Figure 7.4.

The core of the science of grassland improvement will always be situated on the right-hand side of Figure 7.4, since technological innovations in farm practice depend mainly upon advances in the component scientific disciplines. The challenge to grassland scientists is to ensure that communication between interested groups is sufficiently strong to shape the nature of research outputs and to recognize the relevance of new advances so that their adaptation may result in a utility in farming and in resource protection. There is a concomitant need to recognize the role of farmers in generating many of the technological innovations which were subsequently refined by science in dialogue with practice (for example, aerial seeding of forage legumes).

The identification, breeding or creation of new germplasm with, for instance, disease resistance, will only benefit farming as the technology of successful seed production and marketing is developed, reliable techniques of pasture establishment become available, environmental adaptation of the variety is defined and the qualities of the plant become attractive to farmers. The physiological understanding of plant responses to defoliation

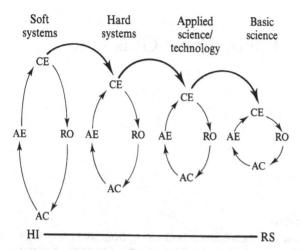

Figure 7.4. The holistic integration (HI) – reductionist science (RS) axis: a spiral of interconnected learning cycles (CE, concrete experience; RO, reflective observation; AC, abstract conceptualization; AE, active experimentation. (From Wilson & Morren, 1990.)

and of the control of plant persistence and replacement needs to be translated as a component of a hard system SR model. In this model the desired optimization of the goals of the landholder with respect to soil conservation, retention of biodiversity, animal performance, acceptance of risk, and current and future economic return are realized through the weighting given to each.

The changes that are sought revolve about the restoration of degraded grasslands, the use of forages in sustaining cropping systems, and the enhanced long-term efficiency of grasslands in meeting global needs for the production of food and fibre, the sequestering of carbon, the protection of landscape, and the viability of communities. These changes will evolve as value is accorded to the activity and beliefs of all the participants and as processes of continuing learning are promoted.

8

Appendix. The International Grassland Congresses

8.1 Locations and attendance*

The International Grassland Congress first met in Germany from 20–31 May 1927. The principal participants were 16 scientists from Austria, Denmark, Finland, Germany, Norway, Sweden and Switzerland, who assembled in Bremen and made a study tour through north west Germany, visiting Emden, Berlin and Dortmund before taking the train to Leipzig. Here, there were two days of scientific discussion at the zoo, revisited subsequently as the site of the 50th anniversary XIII Congress in 1977. The Congress, under the presidency of Prof. A. Falke of Leipzig, had a further study tour through grassland production sites in Saxony before dissolving at Dresden.

The second Congress, which met under the presidency of Dr A. Elofson of Upsala in Sweden and Denmark in 1930, was larger, 58 participants from 13 countries (including Canada). The third Congress, in Switzerland in 1933 with Prof. A. Volkart of Zurich as president, had scientists from Turkey and South Africa present but it was not until the IV Congress in 1937 at Aberystwyth, UK, that the meeting could claim a global constituency. There were some 365 participants from 37 countries; all 11 regions of the world as defined by the 1977 International Grassland Congress Constitution were represented, with the exception of the Middle East (Figure 8.1). The leadership of R.G. Stapledon of the Welsh Plant Breeding Station was pre-eminent. At this meeting it was agreed that the funds of the International Grassland Congress Association be banked in Germany and that the next Congress be held in the Netherlands in 1940. The intervention of the Second World War delayed the V Congress until 1949 and the funds of the Association were not recovered.

The VI International Grassland Congress, held at State College, Pennsylvania, USA in 1952, built on the European foundations of the movement to enlarge its scientific content and global representation, and accorded a new maturity. The world regions with an established history of grassland research (North America, Western Europe, Australia and New Zealand) accounted for 75% of the 271 scientific papers presented, and the participation of other regions of the world increased to 25%, which is the average figure for all 17 Congresses and also for the most recent 1993 Congress (Figure 8.1).

*Material in this section is drawn from the proceedings of the 17 International Grassland Congresses.

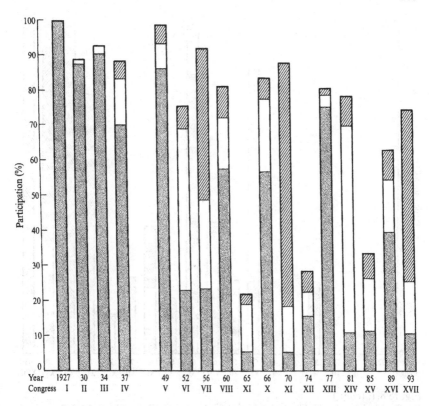

Figure 8.1. Percent attendance at International Grassland Congresses by region
(a) Western Europe ▨, North America □, Oceania ▧

The location of subsequent Congresses usually alternated between continents: America (2), Oceania (3), Asia (1), Europe (5) but no Congress has been held in Africa. The VII Congress in 1956 at Palmerston North, New Zealand had a restricted representation (Figure 8.2) but the VIII Congress at Reading, UK (591 participants from 53 countries) indicated the continued strength of grassland science. At this Congress an Executive Committee representative of eight regions of the world was elected, with a rotating membership so that members would serve for a period covering the two intervals between three Congresses. The full membership of the Congress voted for the IX Congress venue of Brazil, and this was held in 1965 at Sao Paulo, the first congress to be located in a tropical country. In Brazil it was decided that the venue of the XI Congress would be Australia (118 votes, Canada 63 votes, USSR 63 votes). The X Congress moved closer to the Arctic Circle in 1966 at Helsinki, Finland where USSR (128 votes) was selected over Canada (108 votes) for the XII Congress site. The designation of Executive Committee was altered to that of a 'Continuing Committee', which was *inter alia* given the responsibility 'to select the host country for the forthcoming Congress and to announce the name of that host country at the immediate Congress'. The XI Congress was mounted in 1970 at Surfers' Paradise, Queensland, Australia.

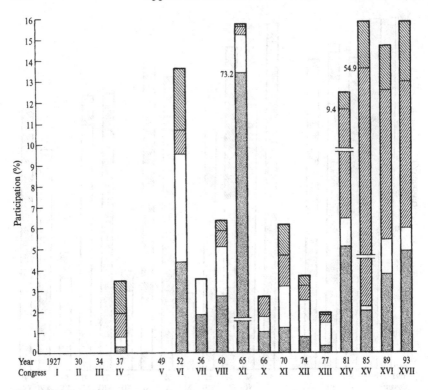

Figure 8.1. (*cont.*)
(b) South America ▨, Central America □, East Asia ▨, Southern Asia ▧

The question of the venue of the XIII Congress aroused controversy at the XII Congress in 1974 in Moscow. The Continuing Committee, empowered by the Constitution adopted in 1966 at Helsinki, determined the Republic of Ireland as the venue. This decision was challenged by the Host Committee in Moscow who put the question to a free vote of full Congress members, of whom 64% were from the northern Eurasia region. This resulted in a decision for the XIII Congress to be held in 1977 at Leipzig, German Democratic Republic. (It is reported that at this meeting a USSR official on the platform turned to R.J. Bula, the North American proxy delegate on the Continuing Committee, when the vote was announced and asked 'So how do you enjoy democracy?'). A further resolution led to the promulgation of a new constitution which was adopted at the Leipzig Congress and which reaffirmed the power of the Continuing Committee to determine future venues, subject to a Congress one country – one vote procedure in the event of a disagreement in the Continuing Committee.

S.C. Pandeya, the outgoing chairman of the Continuing Committee, had expected to invite the XIV Congress to India, but the defeat of the Gandhi government by Mr Desai put paid to this proposition and no invitation from other countries was forthcoming. Canada had previously sought to host congresses but 1977 was not a propitious time to find support. The American Forage and Grassland

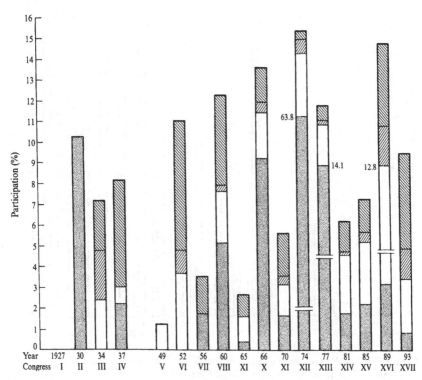

Figure 8.1. (*cont.*)
(c) Northern Eurasia ▨, Mediterranean ▨, Middle East □, Africa ▨

Council, led by R.F. Barnes and J.E. Baylor, ventured in faith and the XIV Congress at Lexington, Kentucky, USA resulted in 1981. The XV Congress in 1985 was the first Congress to be held in Asia and at Kyoto, Japan, a large delegation of scientists from China attended for the first time.

Previous Congresses in Europe were held in cold northern latitudes and the XVI Congress in 1989 at Nice, France was the first Mediterranean location and attracted a higher proportion of participants (13%) from the designated Mediterranean region countries, whilst France provided 24% of the attendance. The XVII Congress in 1993 was unusual in that it arose from the joint invitation of New Zealand and Australia, and its locations in Palmerston North, Hamilton, and Christchurch, New Zealand and Rockhampton, Queensland provided a range of ecological conditions including both temperate and tropical pastures. This was the largest and most representative Congress (Figure 8.2). The scientific contribution and leadership of indigenous participants from the developing countries increased substantially at the XVI and XVII Congresses; in the early Congresses their rather meagre representation often arose from expatriates working in those countries. The invitation of Canada to host the XVIII Congress in 1997 was accepted by the 1993 Continuing Committee.

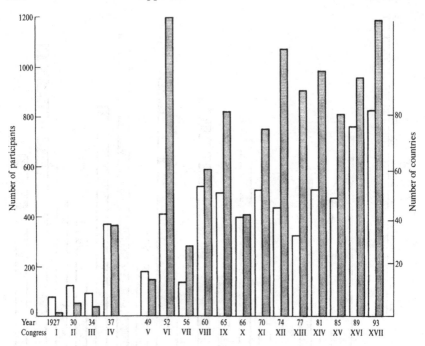

Figure 8.2. Total attendance of participants ▓ and number of countries represented □ at International Grassland Congresses.

8.2 The International Rangeland Congress

The management of rangelands, focused on natural pastures in the arid and semi-arid zones, has always been a topic at International Grassland Congresses and has received varying attention. However, some scientists working in this general area considered there was a need for a separate international meeting directed to developing a better science of the manipulation, improvement and utilization of rangelands. This was exacerbated in the USA by the dichotomy of effort between members of the Society of Range Management and of the American Forage and Grassland Council, whose primary interests were in planted grasslands. The decision to form a separate organization which would mount International Rangeland Congresses was further stimulated by the decision at the XII International Grassland Congress in 1974 to reject the Continuing Committee's acceptance of the Republic of Ireland as the venue for the XIII Congress and to retain the Congress in what was perceived as the Eastern Bloc venue of the German Democratic Republic.

The first International Rangeland Congress was held in 1978 at Denver, Colorado, USA and was succeeded in 1984 by the second Congress at Adelaide, Australia. This was attended by 499 participants from 42 countries; of these 79% came from Oceania, North America and Western Europe. The third Congress in 1988 met in New Delhi, India, and the fourth Congress was held in 1991 at Montpellier, France, whilst the fifth Congress returned to the USA in Utah. Reciprocal representation on the two Congress Continuing Committees was arranged from 1981.

Table 8.1 *Themes represented at International Grassland Congresses (percentage of papers with main theme)*

Subject theme	Year of Congress (Congress number)					Mean
	1937 (IV)	1952 (VI)	1966 (X)	1981 (XIV)	1993 (XVII)	
1. Styles of grassland improvement; regional themes	23	18	10	9	16	15
2. Plant genetic base	23	20	19	21	25	22
3. Edaphic constraints	19	13	14	14	11	14
4. Perspectives from plant physiology	9	9	17	15	14	13
5. Ecology of grasslands	6	10	4	6	7	7
6. Grazing systems	4	4	8	8	8	6
7. Nutritive value	4	7	15	10	7	9
8. Continuity of forage supply	7	11	8	10	5	8
9. Systems approach	–	0.4	1	3	3	1
10. Socio-economic perspectives	1	3	3	4	4	3
11. Miscellaneous	3	6	3	1	2	3
No. entries	69	256	220	480	943	

Source: IGC (1937, 1952, 1966, 1981, 1983, 1993).

8.3 Changes in the balance of themes

8.3.1 Overview

The changing themes which have occupied scientists at International Grassland Congresses were analysed by identifying 110 topics grouped within ten main themes, and additionally including four miscellaneous themes: synoptic papers, biometrics, agricultural engineering, and animal production not specifically related to grassland improvement (Table 8.1). Papers presented at Congresses were allocated to each sub-theme according to its major content; this was not necessarily the theme of the Congress session to which it may have been allocated for convenience.

The content of five Congresses which were held at a mean interval of 14 years from 1937 to 1993 was studied; 1937 was chosen as the first Congress which could claim a good international status. All five Congresses accepted voluntary papers and were held in regions with a history of research in grassland science.

This analysis revealed a considerable homeostasis of disciplinary content during 56 years. The science of grassland improvement has relied first on an interest in its plant genetic base, and plant genetics, plant physiology, plant ecology and soil science contributed 52–57% of the subject matter at all five Congresses. Animal nutrition and systems of animal production arising from study of the animal–plant–soil interface were the other key preoccupations of grassland scientists, whilst environmental science, systems theory and socio-economic perspectives emerged with more force in recent Congresses.

8.3.2 The 1937–1952 period

The general theme of the first subject in Table 8.1, which was designated as styles of grassland development, included the papers with general or integrative themes which were insufficiently specific to be allocated elsewhere and whose main interest was regional or local. These constituted 19% of papers in 1937 and mainly dealt with humid or subhumid temperate grasslands; in 1952 this category decreased to 12%, with a predominance of non-specific tropical papers. The balance of content focused on intensity of land use, tree crops with pastures, leys and turf.

The plant genetic basis for grassland development in 1937 was oriented to evaluation of and selection within improved species; in 1952 there were more papers on hybridization, induced polyploidy, disease resistance and certification of seed for varietal purity. Edaphic constraints on grassland development in 1952 were defined less in terms of general fertilizer needs and responses and more in terms of specific nutrients, including S and soil toxicities; soil conservation and watershed management became significant emphases. More interest in the physiology of flowering and seed production emerged, whilst in plant succession, the control of weed and shrub encroachment and the production of inventories of grassland resources were of significant interest in grassland ecology.

In the 1952 Proceedings studies of selective grazing and foraging strategy, stocking rate and forage allowance, and the methodology of grazing experiments appeared. More sophisticated approaches to nutritive value of forage were evident in the attention to energy value, digestibility and intake, mineral content, and anti-quality factors. Continuity of forage supply was addressed through irrigation and techniques of crop processing, which were especially dependent upon innovations in agricultural engineering. Characterization of climate emerged as a topic, as did the transfer of technology to farmers.

8.3.3 The 1966 Congress

The trend to fewer general papers of regional interest continued, especially, at this Finnish venue, with respect to tropical grasslands. Papers dealing with specialist techniques of plant breeding such as induced polyploidy were again presented. The intensive use of fertilizer N was a new emphasis and there were 19 papers on this topic. Plant physiology was accorded greater importance through papers on growth analysis, tillering, plant response to defoliation, and the role of carbohydrate 'reserves', but there were fewer papers on plant ecology.

Stocking method, stocking rate and forage allowance were further addressed, together with the spatial transfer of nutrients under grazing. Mixed grazing and the innovative choice of animal species were canvassed. Nutritive value received increased attention relative to 1952, especially in relation to forage intake, digestibility and anti-quality factors. Animal responses to systems of forage conservation were described and systems modelling in grassland research appeared as a topic (Table 8.1).

8.3.4 The 1981 Congress

This Congress was marked by considerable advances in tropical pasture science, and 133 of the 480 papers presented bore directly on grassland development in the tropics and subtropics, mainly in specialist areas. Styles of grassland development embraced interest in the intensity of land use, integration of land classes, deforestation and woodland management, long-term trends in production, the use of shrub legumes and intercropping.

Wide approaches to the improvement of the plant genetic base were enunciated which displayed increased emphasis on species evaluation, the conservation of germplasm, and the identification of elite material, whilst *in vitro* embryo culture signalled the nascency of molecular biology.

Scientists at all five Congresses emphasized the role of legumes and of biological N fixation in grassland production; associative mechanisms of N fixation were mentioned at the 1981 Congress, and soil N, together with nutrient cycling, stream pollution, soil toxicities and salinity received increased attention. Perspectives from plant physiology incorporated more interest in pathways of photosynthesis, efficiency of conversion of radiation, moisture use, stress resistance, growth regulators and the understanding of constraints to pasture establishment. The dynamics of change in plant communities, the role of fire, and the control of shrub encroachment figured in grassland ecology, and some 91 papers were directed to the conservation and improvement of natural grasslands.

The influence of grazing on the balance of legumes and grasses and studies of foraging strategies figured in the 1981 Congress. The effect of endophytes, the potential of growth regulators and of chemical processing of crop materials were canvassed. Modelling of grassland systems and the development of decision support systems emerged as strong emphases, whilst technology transfer and the development of the human skills base in grassland science were accorded more significance.

8.3.5 The 1993 Congress

A wider series of topics was structured in depth at the 1993 Congress than had occurred previously. Environmental science was a strong feature of the Congress and the fashionable term 'sustainable development' was explored in its various facets: the properties of systems of land use of varying intensity; tree crops with pastures; alley farming; the role of leys; relict areas; deforestation; and woodland management. Atmospheric pollution and global warming, stream pollution, nutrient leaching and nutrient cycling were components of the agenda, whilst a recurrence of interest in organic matter and soil biological activity reinforced these trends.

Studies of the genetic basis of grassland improvement included more attention to the definition of criteria of merit and of disease resistance, and the rise of genetic engineering and of molecular biology in the allocation of research resources was evident. Many of the themes previously attacked in plant physiology continued from 1981, with more attention to the control of flowering and the processes of seed production. In grassland ecology the dynamics of change in plant communities, the utility of state-and-transition models and the use of remote sensing in producing inventories and current assessments of grassland resources figured strongly.

The perennial themes within the concepts of nutritive value, the devising of grazing systems and the maintenance of continuity of forage supply were elaborated further but in a new context of their description within systems theory.

The socio-economic perspectives which emerged at the 1981 Congress were enlarged by reference to social equity in grassland development, the participation of farmers in grassland research, and to the larger canvases of institutional policies with respect to resource transfer and international trade. The role of Grassland Congresses in assisting scientists working in specialist areas to conceptualize their work in wider contexts was well fulfilled at this Congress.

References

Acocks, J.P.H. (1953). *Veld Types of South Africa*. Pretoria: Government Printer.
Acocks, J.P.H. (1966). Non-selective grazing as a means of veld reclamation. *Proceedings Grassland Society of Southern Africa*, 1, 33–9.
Adams, W.E., White, A.W. Jr. & Dawson, R.N. (1967). Influence of lime sources and rates on 'Coastal' bermuda grass production, soil profile reaction, exchangeable Ca and Mg. *Agronomy Journal*, 59, 147–9.
Adegbola, A. (1966). Preliminary observations on the reserve carbohydrate and regrowth potential of tropical grasses. In *Proceedings X International Grassland Congress*, pp. 933–6.
Adegbola, A.A. & McKell, C.M. (1966). Effect of nitrogen fertilization on the carbohydrate content of coastal bermuda grass. *Agronomy Journal*, 58, 60–64.
Adjei, M.B., Mislevy, P. & Ward, C.Y. (1980). Response of tropical grasses to stocking rate. *Agronomy Journal*, 72, 863–8.
Aii, T. & Stobbs, the late T.H. (1980). Solubility of the protein of tropical pasture species and the rate of its digestion in the rumen. *Animal Feed Science and Technology*, 5, 183–92.
Alberda, T. (1966). The influence of reserve substances on dry-matter production after defoliation. In *Proceedings X International Grassland Congress*, pp. 140–7.
Albertson, F.W., Riegel, A. & Launchbauch, J.L. (1953). Effects of different intensities of clipping on short grasses in west-central Kansas. *Ecology*, 34, 1–20.
Alva, A.K., Blamey, F.P.C., Edwards, D.G. & Asher, C.J. (1986). An evaluation of aluminium indices to predict aluminium toxicity to plants grown in nutrient solutions. *Communications in Soil Science and Plant Analysis*, 17, 1271–80.
Anderson, A.J. (1956). Effects of fertilizer treatments on pasture growth. In *Proceedings VII International Grassland Congress*, pp. 323–33.
Anderson, A.J. & Arnot, R.H. (1953). Fertilizer studies on basaltic red loam from the Lismore district, New South Wales. *Australian Journal of Agricultural Research*, 4, 29–43.
Anderson, A.J. & McLachlan, K.D. (1951). The residual effect of phosphorus on soil fertility and pasture development on acid soils. *Australian Journal of Agricultural Research*, 2, 377–400.
Andreyev, N.G. & Savitskaya, V.A. (1966). Effect of different rates of fertilizers on yields and chemical composition of *Bromus inermis* leys. In *Proceedings X International Grassland Congress*, pp. 279–82.
Archbold, H.K. (1945). Some factors concerned in the process of starch storage in the barley grain. *Nature*, 156, 70–3.

Archer, S. (1993a). Climate change and grasslands: a life-zone and biota perspective. In *Proceedings XVII International Grassland Congress*, pp. 1061–7.

Archer, S. (1993b). Vegetation dynamics in changing environments. *The Rangeland Journal*, **15**, 104–16.

Archer, S., Schimel, D.S. & Holland, E.A. (1995). Mechanisms of shrubland expansion: land use, climate or CO_2? *Climate Change*, **29**, 91–9.

Argel, P.J. & Humphreys, L.R. (1983). Environmental effects on seed development and hardseededness in *Stylosanthes hamata* cv. Verano. III. Storage humidity and seed characteristics. *Australian Journal of Agricultural Research*, **34**, 279–87.

Arora, R.K. & Chandel, K.P.S. (1972). Botanical source areas of wild herbage legumes in India. *Tropical Grasslands*, **6**, 213–21.

Ash, A.J. & McIvor, J.G. (1995). Land condition in the tropical tallgrass pasture lands. 2. Effects on herbage quality and nutrient uptake. *The Rangeland Journal*, **17**, 86–98.

Ash, A.J., McIvor, J.G., Corfield, J.P. & Winter, W.H. (1995). How land condition alters plant–animal relationships in Australia's tropical rangelands. *Agriculture, Ecosystems and Environment*, **56**, 77–9.

Ash, A.J., Prinsen, J.H., Myles, D.J. & Hendricksen, R.E. (1982). Short-term effects of burning native pasture in spring on herbage and animal production in south-east Queensland. In *Proceedings of the Australian Society of Animal Production*, **14**, 377–80.

Ashby, J.A. (1990). *Evaluating Technology with Farmers: A Handbook*. Cali, Colombia: CIAT.

Austin, R.B. (1978). Maximum yields of cereals as determined by plant type and environment. In *Maximising Yields of Crops*, pp. 18–24. London: HMSO.

Ayanaba, A., Tuckwell, S.B. & Jenkinson, D.S. (1976). The effects of clearing and cropping on the organic reserves and biomass of tropical forest soils. *Soil Biology and Biochemistry*, **8**, 519–25.

Bacon, C.W. (1994). Fungal endophytes, other fungi, and their metabolites as extrinsic factors of grass quality. In *Forage Quality, Evaluation and Utilization*, ed. G.C. Fahey Jr, M. Collins, D.R. Mertens & L.E. Moser, pp. 318–366. Madison, WI: American Society of Agronomy.

Bacon, C.W., Porter, J.K., Robbins, J.D. & Luttrell, E.S. (1977). *Epichloe typhina* from toxic tall fescue grasses. *Applied Environmental Microbiology*, **34**, 576–81.

Baker, H.K. (1966). The experimental development of systems of beef production from grassland. In *Proceedings X International Grassland Congress*, pp. 483–7.

Baker, R.D. (1978). Beef cattle at grass: intake and production. In *Proceedings British Grassland Society Winter Meeting*, pp. 2.1–2.7.

Balfourier, F., Charmet, G. & Grand-Ravel, C. (1993). Use of ecogeographical factors for sampling a core collection of perennial ryegrass. In *Proceedings XVII International Grassland Congress*, pp. 228–30.

Barclay, P.C. (1960). Breeding for improved winter pasture production in New Zealand. In *Proceedings VIII International Grassland Congress*, pp. 326–30.

Barnes, D.L. (1961). Residual effects of cutting – frequency and fertilizing with nitrogen on root and shoot growth, and the available carbohydrate and nitrogen content of the roots of Sabi Panicum (*Panicum maximum* Jacq.). *Rhodesia Agricultural Journal*, **58**, 365–9.

Barnes, D.L. (1965). The effects of frequency of burning and mattocking on the control of coppice in the Marandellas sandveld. *Rhodesian Journal of Agricultural Research*, **3**, 55–68.

Barnes, D.L. (1989). Reaction of three veld grasses to different schedules of grazing and resting. 2. Residual effects on vigour. *South African Journal of Plant and Soil*, **6**, 8–13.

Barnes, D.L. (1992). A critical analysis of veld management recommendations for sourveld in the south-eastern Transvaal. *Journal of the Grassland Society of Southern Africa*, **9**, 126–34.

Barnes, D.L. & Hava, K. (1963). Effects of cutting on seasonal changes in the roots of Sabi panicum (*Panicum maximum* Jacq.). *Rhodesia Journal of Agricultural Research*, **1**, 107–10.

Barnes, J.I. (1993). Economic and ecological features of livestock and wildlife utilisation in Africa. In *Proceedings XVII International Grassland Congress*, pp. 2085–91.

Barnes, R.F., Muller, L.D., Bauman, L.F. & Colenbrander, V.F. (1971). *In vitro* dry matter disappearance of brown midrib mutants of maize (*Zea mays* L.). *Journal of Animal Science*, **33**, 881–4.

Barry, T.N. & Blaney, B.J. (1987). Secondary compounds of forages. In *The Nutrition of Herbivores*, ed. J.B. Hacker & J.H. Ternouth, pp. 91–119. Sydney: Academic Press.

Barthram, G.T. & Grant, S.A. (1984). Defoliation of ryegrass-dominated swards by sheep. *Grass and Forage Science*, **39**, 211–19.

Barthram, G.T., Grant, S.A. & Elston, D.A. (1992). The effects of sward height and nitrogen fertilizer application on changes in sward composition white clover growth and the stock carrying capacity of an upland perennial ryegrass white clover sward grazed by sheep for four years. *Grass and Forage Science*, **47**, 326–41.

Bashan, Y. (1993). Potencial use of *Azospirillum* as biofertilizer. *Turrialba*, **43**, 286–91.

Bashaw, E.C., Voigt, P.W. & Burson, B.L. (1983). Breeding challenges in apomictic warm-season grasses. In *Proceedings XIV International Grassland Congress*, pp. 179–81.

Bastin, G.W., Pickup, G., Chewings, V.H. & Pearce, G. (1993). Land degradation assessment in central Australia using a grazing gradient method. *The Rangeland Journal*, **15**, 190–216.

Bawden, R.J., Ison, R.L., Macadam, R.D., Packham, R.G. & Valentine, I. (1985). A research paradigm for systems agriculture. In *Agricultural Systems Research for Developing Countries*, ed. J.V. Remenyi, pp. 31–42. Canberra: ACIAR Proceedings 11.

Bean, E.W. (1972). Clonal evaluation for increased seed production in two species of forage grasses, *Festuca arundinacea* Schreb. and *Phleum pratense* L. *Euphytica*, **21**, 377–83.

Beaty, E.R., Sampaio, E.V.S.B., Ashley, D.A. & Brown, R.H. (1974). Partitioning and translocation of ^{14}C photosynthate by Bahia grass (*Paspalum notatum* Flugge). In *Proceedings XII International Grassland Congress*, pp. 259–67.

Beever, D.E. (1993). Ruminant animal production from forages: present position and future opportunities. In *Proceedings XVII International Grassland Congress*, pp. 535–41.

Begon, M., Harper, J. & Townsend, C.R. (1986). *Ecology, Individuals, Populations and Communities*. Oxford: Blackwell Scientific Publications.

Bethlenfalvay, G.J., Ulrich, J.M. & Brown, M.S. (1985). Plant response to mycorrhizal fungi: host, endophyte and soil effects. *Soil Science Society of America Journal*, **49**, 1164–8.

Betteridge, K., Ulyatt, M.J., Knapp, J. & Baldwin, R.L. (1993). Methane production by New Zealand ruminants. In *Proceedings XVII International Grassland Congress*, pp. 1199–1200.

Biggs, S.D. (1995). Farming systems research and rural poverty: relationships between context and content. *Agricultural Systems*, **47**, 161–74.

Binnie, R.C. & Chestnutt, D.M.B. (1994). Effect of continuous stocking by sheep at four sward heights on herbage mass, herbage quality and tissue turnover on grass/clover and nitrogen-fertilized grass swards. *Grass and Forage Science*, **49**, 192–202.

Bishop, H.G., Pengelly, B.C. & Ludke, D.H. (1988). Classification and description of a collection of the legume genus *Aeschynomene*. *Tropical Grasslands*, **22**, 160–75.

Blackman, G.E. (1936). The influence of temperature and available nitrogen supply on the growth of pastures in spring. *Journal of Agricultural Science*, **26**, 620–47.

Blamey, F.P.C., Asher, C.J., Kerven, G.L. & Edwards, D.G. (1993). Factors affecting aluminium sorption by calcium pectate. *Plant and Soil*, **149**, 87–94.

Blaser, R.E., Brown, R.H. & Bryant, H.T. (1966). The relationship between carbohydrate accumulation and growth of grasses under different microclimates. In *Proceedings X International Grassland Congress*, pp. 147–50.

Blunt, C.G. & Humphreys, L.R. (1970). Phosphate response of mixed swards at Mt Cotton, South-eastern Queensland. *Australian Journal of Experimental Agriculture and Animal Husbandry*, **10**, 431–41.

Boddey, R.M. & Döbereiner, J. (1988). Nitrogen fixation associated with grasses and cereals: recent results and perspectives for future research. *Plant and Soil*, **108**, 53–65.

Bogdan, A.C. (1977). *Tropical pastures and fodder plants*. London: Longman.

Böhnert, E., Lascano, C. & Weniger, J.H. (1985). Botanical and chemical composition of the diet selected by fistulated steers under grazing on improved grass-legume pastures in the tropical savannas of Colombia. I. Botanical composition of forage available and selected. *Zonderdruck aus Zeitschrift für Tierzüchtung und Züchtungsbiologie*, **102**, 385–94.

Böhnert, E., Lascano, C. & Weniger, J.H. (1986). Botanical and chemical composition of the diet selected by fistulated steers under grazing on improved grass-legume pastures in the tropical savannas of Colombia. II. Chemical composition of forage available and selected. *Zonderdruck aus Zeitschrift für Tierzüchtung und Züchtungsbiologie*, **103**, 69–79.

Bommer, D.F.R. (1966). Influence of cutting frequency and nitrogen level on the carbohydrate reserves of three grass species. In *Proceedings X International Grassland Congress*, pp. 156–60.

Boonman, J.G. (1981). Tropical pasture management in Kenya. In *XIV International Grassland Congress, Summaries of Papers*, p. 210. Lexington, KY: University of Kentucky.

Booysen, P. de V., Tainton, N.M. & Scott, J.D. (1963). Shoot apex development in grasses and its importance in grassland management. *Herbage Abstracts*, **33**, 209–13.

Bouton, J.H., Gates, R.W., Belesky, D.P. & Owsley, M. (1993a). Yield and persistence of tall fescue in the southeastern Coastal Plain after removal of its endophyte. *Agronomy Journal*, **85**, 52–5.

Bouton, J.H., Smith, S.R. Jr, Hoveland, C.S. & McCann, M.A. (1993b). Evaluation of grazing-tolerant alfalfa cultivars. In *Proceedings XVII International Grassland Congress*, pp. 416–18.

Bowen, E.J. & Rickert, K.G. (1979). Beef production from native pastures sown to fine-stem stylo in the Burnett region of south-eastern Queensland. *Australian Journal of Experimental Agriculture and Animal Husbandry*, **19**, 140–9.

Bowen, G.D. & Kennedy, M.M. (1959). Effect of high soil temperatures on *Rhizobium* spp. *Queensland Journal of Agricultural Science*, **16**, 177–97.

Braithwaite, K.S., Irwin, J.A.G. & Manners, J.M. (1990). Restriction fragment length polymorphisms in *Colletotrichum gloeosporioides* infecting *Stylosanthes* spp. in Australia. *Mycological Research*, **94**, 1129–37.

Branson, F.A. (1953). Two new factors affecting resistance of grasses to grazing. *Journal of Range Management*, **6**, 165–71.

Branson, F.A. (1956). Quantitative effects of clipping treatments on five range grasses. *Journal of Range Management*, **9**, 86–8.

Bray, R.A. (1994). The *Leucaena* psyllid. In *Forage Tree Legumes in Tropical Agriculture*, ed. R.C. Gutteridge & H.M. Shelton, pp. 283–91. Wallingford, UK: CAB International.

Bray, R.A. & Hutton, E.M. (1976). Plant breeding and genetics. In *Tropical Pasture Research. Principles and Methods*, ed. N.H. Shaw & W.W. Bryan, pp. 338–53. Farnham Royal, UK: CAB.

Bray, R.A. & Sands, D.P.A. (1987). Arrival of the *Leucaena* psyllid in Australia: impact, dispersal and natural enemies. *Leucaena Research Reports*, **7**(2), 61–5.

Bredon, R.M. & Horrell, C.R. (1963). Management studies with *Panicum maximum* in Uganda. II. The effect of cutting interval and nitrogen fertiliser on chemical composition and nutritive value. *Empire Journal of Experimental Agriculture*, **31**, 343–50.

Breese, E.L. (1983). Exploitation of the genetic resource through breeding: *Lolium* species. In *Genetic Resources of Forage Plants*, ed. J.G. McIvor & R.A. Bray, pp. 275–88. Melbourne: CSIRO.

Breese, E.L. & Davies, W.E. (1970). Herbage plant breeding. In *Jubilee Report of the Welsh Plant Breeding Station 1919–1969*, pp. 11–47. Aberystwyth: Welsh Plant Breeding Station.

Brewbaker, J.L. & Atwood, S.A. (1952). Incompatibility alleles in polyploids. In *Proceedings VI International Grassland Congress*, pp. 267–72.

Brewbaker, J.L. & Sorensson, C.T. (1990). New tree crops from interspecific *Leucaena* hybrids. In *Advances in New Crops*, ed. J. Janick & J.E. Simon, pp. 283–9. Portland, Oregon: Timber Press.

Breymeyer, A.I. & Van Dyne, G.M. (eds) (1980). *Grasslands, Systems Analysis and Man*. IBP 19. Cambridge: Cambridge University Press.

Bridge, B.J., Mott, J.J., Winter, W.H. & Hartigan, R.J. (1983). Improvement in soil structure resulting from sown pastures on degraded areas in the dry savanna woodlands of northern Australia. *Australian Journal of Soil Research*, **21**, 83–90.

Briggs Myers, I. & McCauley, M.H. (1985). *Manual: A Guide to the Development and Use of the Myers–Briggs Type Indicator*. Palo Alto, California: Consulting Psychological Press.

Briske, D.D. & Silvertown, J.W. (1993). Plant demography and grassland community balance: the contribution of population regulation mechanisms. In *Proceedings of XVII International Grassland Congress*, pp. 291–8.

Brock, R.D. (1971). The role of induced mutations in plant improvement. *Radiation Botany*, **11**, 181–96.

Brockington, N.R. (1974). A systems approach to grassland research. In *Silver Jubilee Report*, ed. C.R.W. Spedding & R.D. Williams, pp. 179–84. Hurley, UK: Grassland Research Institute.

Brockwell, J. (1962). Studies on seed pelleting as an aid to legume seed inoculation. 1. Coating materials, adhesives, and methods of inoculation. *Australian Journal of Agricultural Research*, **13**, 638–49.

Broderick, G.A. (1994). Quantifying forage protein quality. In *Forage Quality, Evaluation and Utilization*, ed. G.C. Fahey Jr, M. Collins, D.R. Mertens & L.E. Moser, pp. 200–28. Madison, WI: American Society of Agronomy.

Brougham, R.W. (1956). Effect of intensity of defoliation on regrowth of pasture. *Australian Journal of Agricultural Research*, **7**, 377–87.

Brougham, R.W. (1958). Interception of light by the foliage of pure and mixed stands of pasture plants. *Australian Journal of Agricultural Research*, **9**, 39–52.

Brougham, R.W. (1970). The approach to grassland as a soil–plant–animal system. In *Proceedings XI International Grassland Congress*, pp. A120–6.

Brown, B.A. (1937). The effects of fertilizers on the soil, the botanical and chemical composition of the herbage and the seasonal and total production of grassland in Connecticut. In *Report IV International Grassland Congress*, pp. 313–17.

Brown, R.H. & Blaser, R.E. (1968). Leaf area index in pasture growth. *Herbage Abstracts*, **38**, 1–9.

Bryan, W.W. & Evans, T.R. (1973). Effects of soils, fertilizers and stocking rates on pastures and beef production on the wallum of southeast Queensland 1. Botanical composition and chemical effects on plants and soils. *Australian Journal of Experimental Agriculture and Animal Husbandry*, **13**, 516–29.

Bryant, F.C., Dahl, B.E., Petit, R.D. & Britton, E.M. (1989). Does short-duration grazing work in arid and semi-arid regions? *Journal of Soil and Water Conservation*, **44**, 290–6.

Bucher, H.P., Mächler, F. & Nösberger, J. (1987). Sink control of assimilate partitioning in meadow fescue (*Festuca pratensis* Huds.). *Journal of Plant Physiology*, **129**, 469–77.

Budelman, A. (1988). The decomposition of the leaf mulches of *Leucaena leucocephala*, *Gliricidia sepium* and *Flemingia macrophylla* under humid tropical conditions. *Agroforestry Systems*, **7**, 33–45.

Bughrara, S.S. & Sleper, D.A. (1986). Digestion of several temperate forage species by a prepared cellulase solution. *Agronomy Journal*, **78**, 94–8.

Bunting, A.H. (1983). Review and prospect. In *Genetic Resources of Forage Plants*, ed. J.G. McIvor & R.A. Bray, pp. 313–21. Melbourne: CSIRO.

Burke, M.J., Gusta, L.V., Quamme, H.A., Weiser, C.J. & Li, P.H. (1976). Freezing and injury in plants. *Annual Review of Plant Physiology*, **27**, 507–28.

Burlison, A.J., Hodgson, J. & Illius, A.W. (1991). Sward canopy structure and the bite dimensions and bite weight of grazing sheep. *Grass and Forage Science*, **46**, 29–38.

Burnside, D.G. & Faithfull, E. (1993). Judging range trend: interpretation of rangeland monitoring data drawn from sites in the Western Australian shrublands. *The Rangeland Journal*, **15**, 247–69.

Burrows, W.H. (1995). Greenhouse revisited – land-use change from a Queensland perspective. *Climate Change Newsletter*, **7**(1), 6–7.

Burt, R.L., Edye, L.A., Williams, W.T., Gillard, P., Grof, B., Page, M., Shaw, N.H., Williams, R.J. & Wilson, G.P.M. (1974). Small-sward testing of *Stylosanthes* in northern Australia: preliminary considerations. *Australian Journal of Agricultural Research*, **25**, 559–75.

Burt, R.L., Edye, L.A., Williams, W.T., Grof, B. & Nicholson, C.H.L. (1971). Numerical analysis of variation patterns in the genus *Stylosanthes* as an aid to plant introduction and assessment. *Australian Journal of Agricultural Research*, **22**, 737–57.

Burt, R.L., Reid, R. & Williams, W.T. (1976). Exploration for, and utilization of, collections of tropical pasture legumes. I. The relationship between agronomic performance and climate of origin of introduced *Stylosanthes* spp. *Agro-Ecosystems*, **2**, 293–307.

Burt, R.L. & Williams, W.T. (1975). Plant introduction and the *Stylosanthes* story. *AMRC Review*, **25**, 1–26.

Burton, G.W. (1952). Quantitative inheritance in grasses. In *Proceedings VI International Grassland Congress*, pp. 277–83.

Burton, G.W. (1983). Improving the efficiency of forage-crop breeding. In *Proceedings XIV International Grassland Congress*, pp. 138–40.

Burton, G.W. & Forbes, I. (1960). The genetics and manipulation of obligate apomixis in common bahiagrass (*Paspalum notatum* Flugge.) In *Proceedings VIII International Grassland Congress*, pp. 66–71.

Burton, G.W., Hart, R.H. & Lowrey, R.S. (1967). Improving forage quality in bermudagrass by breeding. *Crop Science*, **7**, 329–32.

Burton, G.W. & Jackson, J.E. (1962). A method for measuring sod reserves. *Agronomy Journal*, **54**, 53–5.

Burton, G.W., Prine, G.M. & Jackson, J.E. (1957). Studies of drought tolerance and water use of several southern grasses. *Agronomy Journal*, **49**, 498–503.

Bushby, H.V.A. (1982). Rhizosphere populations of *Rhizobium* strains and nodulation of *Leucaena leucocephala*. *Australian Journal of Experimental Agriculture and Animal Husbandry*, **22**, 293–8.

Busso, C.A., Richards, J.H. & Chatterton, N.J. (1990). Nonstructural carbohydrates and spring regrowth of two cool-season grasses: interaction of drought and clipping. *Journal of Range Management*, **43**, 336–43.

Butler, G.W. & Bathurst, N.O. (1956). The underground transference of nitrogen from clover to associated grass. In *Proceedings VII International Grassland Congress*, pp. 168–78.

Butler, G.W., Greenwood, R.M. & Soper, K. (1959). Effects of shading and defoliation on the turnover of root and nodule tissue of plants of *Trifolium repens, Trifolium pratense* and *Lotus uliginosus*. *New Zealand Journal of Agricultural Research*, **2**, 415–26.

Buxton, D.R. & Lentz, E.M. (1993). Performance of morphologically diverse orchard grass clones in spaced and sward plantings. *Grass and Forage Science*, **48**, 336–46.

Cadisch, O., Schunke, R.M. & Giller, K.E. (1994). Nitrogen cycling in a pure grass pasture and a grass–legume mixture on a red latosol in Brazil. *Tropical Grasslands*, **28**, 43–52.

Calvin, M. (1962). The path of carbon in photosynthesis. *Science*, **135**, 879.

Cameron, D.F. (1983). To breed or not to breed. In *Genetic Resources of Forage Plants*, ed. J.G. McIvor & R.A. Bray, pp. 237–50. Melbourne: CSIRO.

Cameron, D.F., Bishop, H.G., Mannetje, L.'t, Shaw, N.H., Sillar, D.I. & Staples, I.B. (1977). The influence of flowering time and growth habit on the performance of townsville stylo (*Stylosanthes humilis*) in tropical and sub-tropical Queensland. *Tropical Grasslands*, **11**, 165–75.

Cameron, D.F., Boland, R.A., Chakraborty, S., Jamieson, B. & Irwin, J.A.G. (1993b). Recurrent selection for partial resistance to anthracnose disease in shrubby stylo (*Stylosanthes scabra*). In *Proceedings XVII International Grassland Congress*, pp. 2137–8.

Cameron, D.F., Miller, C.P., Edye, L.A. & Miles, J.W. (1993a). Advances in research and development with *stylosanthes* and other tropical pasture legumes. In *Proceedings XVII International Grassland Congress*, pp. 2109–14.

Campbell, A.G. (1970). Recent advances in the control of facial eczema. In *Proceedings XI International Grassland Congress*, pp. 774–7.

Campbell, B.D. & Grime, J.P. (1993). Prediction of grassland plant responses to global change. In *Proceedings XVII International Grassland Congress*, pp. 1109–18.

Campbell, B. & Stafford Smith, M. (1993). Defining GCTE modelling needs for pastures and rangelands. In *Proceedings XVII International Grassland Congress*, pp. 1249–53.

Caradus, J.R. (1993). Progress in white clover agronomic performance through breeding. In *Proceedings XVII International Grassland Congress*, pp. 396–7.

Cardon, P.V. (1937). Plant breeding in relation to pasture improvement. In *Report IV International Grassland Congress*, pp. 31–9.

Cardon, P.V. (1952). Our concept of grassland agriculture. In *Proceedings VI International Grassland Congress*, pp. 92–9.

Carew, G.W. (1976). Stocking rate as a factor determining profitability of beef production. *Rhodesia Agricultural Journal*, **73**, 111–15.

Carlson, G.E. (1966). Growth of white clover leaves, after leaf removal. In *Proceedings X International Grassland Congress*, pp. 134–6.

Carmi, A. & Koller, D. (1979). Regulation of photosynthetic activity in the primary leaves of bean (*Phaseolus vulgaris* L.) by materials moving in the water-conducting system. *Plant Physiology*, **64**, 285–8.

Carran, R.A. (1993). Response of white clover to soil phosphate levels at different temperatures and carbon dioxide concentrations. In *Proceedings XVII International Grassland Congress*, pp. 1135–7.

Carter, E.D. & Porter, R.G. (1993). Predicting the emergence of annual pasture legumes in cereal–livestock forming systems of South Australia. In *Proceedings XVII International Grassland Congress*, pp. 2193–4.

Carter, J.O. (1994). *Acacia nilotica* – a tree legume out of control. In *Forage Tree Legumes in Tropical Agriculture*, ed. R.C. Gutteridge & H.M. Shelton, pp. 338–51. Wallingford, UK: CAB International.

Cashmore, A.B. (1934). A comparative study of *Lolium perenne* and *Phalaris tuberosa* at varying stages of growth. Melbourne: CSIR Bulletin 81.

Catchpoole, D.W. & Blair, G.J. (1990). Forage tree legumes. I. Productivity and N economy of *Leucaena, Gliricidia, Calliandra* and *Sesbania* and tree/green panic mixtures. *Australian Journal of Agricultural Research*, **41**, 521–30.

Cerri, C.C., Volkoff, B. & Andreaux, F. (1991). Nature and behaviour of organic matter in soils under natural forest, and after deforestation, burning and cultivation, near Manaus. *Forest Ecology and Management*, **38**, 247–57.

Chakraborty, S., Cameron, D.F. & Lupton, J. (1996). Management through understanding – a case history of *Stylosanthes* anthracnose in Australia. In *Pasture and Forage Crop Pathology*, ed. S. Chakraborty, K.T. Leath, R.A. Skipp, G.A. Pederson, R.A. Bray, G.C.M. Latch & F.W. Nutter Jr, pp. 603–19. Madison, WI: American Society of Agronomy.

Chakraborty, S., Pettit, A.N., Boland, R.M. & Cameron, D.F. (1990). I. Epidemiology and inheritance of rate reducing resistance to anthracnose in *S. scabra* (second and third years). In *Biennial Research Report 1988–90*, pp. 126–7. Brisbane: CSIRO Division of Tropical Crops and Pastures.

Chapin III, F.S., Bloom, A.J., Field C.B. & Waring, R.H. (1987). Plant responses to multiple environmental factors. *Bioscience*, 37, 49–57.

Chapman, D.F. & Lemaire, G. (1993). Morphogenetic and structural determinants of plant regrowth after defoliation. In *Proceedings XVII International Grassland Congress*, pp. 95–104.

Chapman, D.F., Robson, M.J. & Snaydon, R.W. (1992). The carbon economy of clonal plants of *Trifolium repens* L. *Journal of Experimental Botany*, **43**, 427–34.

Checkland, P. (1981). *Systems Thinking, Systems Practice*. New York: John Wiley.

Checkland, P.B. & Scholes, J. (1990). *Soft Systems Methodology in Action*. New York: John Wiley.

Cheng, Y-K. (1993). Somatic embryogenesis and plant regeneration from long-term suspension cultures of *Digitaria decumbens*. In *Proceedings XVII International Grassland Congress* , pp. 1038–9.

Chestnutt, D.M.B. (1992). Effect of sward surface height on the performance of ewes and lambs continuously grazed on grass/clover and nitrogen-fertilized grass swards. *Grass and Forage Science*, **47**, 70–80.

Chopping, G.D., Deans, H.D., Sibbick, R., Thurbon, P.N. & Stokoe, J. (1976). Milk production from irrigated nitrogen fertilized Pangola grass. *Proceedings Australian Society of Animal Production*, **11**, 481–4.

Chopping, G.D., Lowe, K.J. & Clarke, L.G. (1983). Irrigation systems. In *Dairy Management in the 80s. Focus on Feeding Seminar*, pp. 109–20. Brisbane: Queensland Department of Primary Industries.

Chopping, G.D., Thurbon, P.N., Moss, R.J. & Stephenson, H. (1982). Winter–spring milk production responses from the annual autumn oversowing of irrigated tropical pastures with rye grass and clovers. *Animal Production in Australia*, **17**, 421–4.

Chowdhury, M.S., Marshall, K.C. & Parker, C.A. (1968). Growth rates of *Rhizobium trifolii* and *Rhizobium lupini* in sterilized soils. *Australian Journal of Agricultural Research*, **19**, 919–25.

Christian, C.S. (1952). Cattle pastures of tropical Australia. In *Proceedings VI International Grassland Congress*, pp. 1534–9.

CIAT. (1976). Pasture utilization. In *Annual Report 1976*, pp. C-32–4. Cali, Colombia: CIAT.

Clark, H. (1993). Relative performance of three cultivars of late-heading perennial ryegrass continuously stocked by ewes and lambs. In *Proceedings XVII International Grassland Congress*, pp. 456–7.

Clatworthy, J.N. (1986). Establishment and yields of pasture legumes under cutting in Zimbabwe. 2. Legumes for ley pasture. *Zimbabwe Journal of Agricultural Research*, **24**, 149–66.

Clausen, J. (1952). New bluegrasses by combining and rearranging genomes of contrasting *Poa* species. In *Proceedings VI International Grassland Congress*, pp. 216–21.

Clement, C.R. (1971). *Residual Effects of Nitrogen and Potassium Applied to Grass*. MAFF Technical Bulletin No. 20, pp. 283–95. London: HMSO.

Clements, F.E. (1916). *Plant Succession: An Analysis of the Development of Vegetation*. Washington: Carnegie Institute Publication 242.

Clements, F.E. (1920). *Plant Indicators. The Relation of Plant Communities to Process and Practice*. Washington: Carnegie Institute Publication 290.

Clements, R.J. (1989). Rates of destruction of growing points of pasture legumes by grazing cattle. In *Proceedings XVI International Grassland Congress*, pp. 1027–8.

Clements, R.J., Hayward, M.D. & Byth, D.E. (1983). Genetic adaptation in pasture plants. In *Genetic Resources of Forage Plants*, ed. J.G. McIvor & R.A. Bray, pp. 101–15. Melbourne: CSIRO.

Clements, R.J. & Ludlow, M.M. (1977). Frost avoidance and frost resistance in *Centrosema virginianum*. *Journal of Applied Science* **14**, 551–6.

Clewett, J.F., Cavaye, J.M., McKeon, G.M., Partridge, I.J. & Scanlan, J.C. (1991). Decision support software as an aid to managing pasture systems. *Tropical Grasslands*, **25**, 159–64.

Clifford, P.E. (1977). Tiller bud suppression in reproductive plants of *Lolium multiflorum* Lam. cv. Westerwoldicum. *Annals of Botany*, **41**, 605–15.

Collins, R.P., Glendining, M.J. & Rhodes, I. (1991). The relationships between stolon characteristics, winter survival and annual yield in white clover (*Trifolium repens* L.). *Grass and Forage Science*, **46**, 51–61.

Collins, R.P. & Rhodes, I. (1989). Yield of white clover populations in mixture with contrasting perennial ryegrasses. *Grass and Forage Science*, **44**, 111–15.

Conniffe, D., Browne, D. & Walshe, M.J. (1970). Experimental design for grazing trials. *Journal of Agricultural Science*, **74**, 339–42.

Connolly, J. (1976). Some comments on the shape of the gain-stocking rate curve. *Journal of Agricultural Science*, **86**, 103–9.

Conway, G.R. (1994). Sustainability in agricultural development: trade-offs between productivity, stability and equitability. *Journal for Farming Systems Research-Extension*, 4(2), 1–14.

Cooper, J.P. (1960). The use of controlled life-cycles in the forage grasses and legumes. *Herbage Abstracts*, **30**, 71–9.

Cooper, J.P. (1964). Leaf development in climatic races of *Lolium* and *Dactylis*. *Journal of Applied Ecology*, **1**, 45–61.

Cooper, J.P. (1966). The significance of genetic variations in light interception and conversion for forage-plant breeding. In *Proceedings X International Grassland Congress*, pp. 715–20.

Cooper, J.P. & Breese, E.L. (1980). Breeding for nutritive quality. *Proceedings of the Nutrition Society*, **39**, 281–6.

Cooper, J.P. & Wilson, D. (1970). Variation in photosynthetic rate in *Lolium*. In *Proceedings XI International Grassland Congress*, pp. 522–7.

Copland, J.W., Djajanegra, A. & Sabrani, M. (1994). *Agroforestry and Animal Production for Human Welfare*. Canberra: ACIAR Proceedings 55.

Coupland, R.T. (1952). Grassland communities of the western Canadian prairies – climax and subclimax. In *Proceedings VI International Grassland Congress*, pp. 625–31.

Coupland, R.T., Skoglund, N.A. & Heard, A.J. (1960). Effects of grazing in the Canadian mixed prairie. In *Proceedings VIII International Grassland Congress*, pp. 213–5.

Cowan, R.T., Byford, I.J.R. & Stobbs, T.H. (1975). Effects of stocking rate and energy supplementation on milk production from tropical grass–legume pasture. *Australian Journal of Experimental Agriculture and Animal Husbandry*, **15**, 740–6.

Cowan, R.T., Davison, T.M. & O'Grady, P. (1977). Influence of level of concentrate feeding on milk production and pasture utilization by Friesian cows grazing tropical grass–legume pasture. *Australian Journal of Experimental Agriculture and Animal Husbandry*, **17**, 373–9.

Cowling, D.W. (1966). The effect of the early application of nitrogenous fertiliser and of the time of cutting in spring on the yield of rye grass/white clover swards. *Journal of Agricultural Science*, **66**, 413–31.

Cowling, D.W. & Clement, C.R. (1974). Aspects of the mineral nutrition of temporary grassland. In *Silver Jubilee Report*, ed. C.R.W. Spedding & R.D. Williams, pp. 81–8. Hurley, UK: Grassland Research Institute.

Cowling, D.W. & Lockyer, D.R. (1967). A comparison of the reaction of different grass species to fertilizer nitrogen and to growth in association with white clover. II. Yield of nitrogen. *Journal of the British Grassland Society*, 22, 53–61.

Cox, P.G. (1996). Some issues in the design of agricultural Decision Support Systems. *Agricultural Systems*, in press.

Cox, P., Parton, K., Shulman, A. & Ridge, P. (1995). On the articulation of simulation and heuristic models of agricultural production systems. In *Proceedings 2nd International Symposium on Systems Approaches for Agricultural Development*, 6–8 December, 1995. Los Baños, Laguna, Philippines: International Rice Research Institute.

Cox, P.G., Shulman, A.D., Ridge, P.E., Foale, M.A. & Garside, A.L. (1996). An interrogative approach to system diagnosis: an invitation to the dance. *Journal of Farming Systems Research and Extension*, in press.

Crampton, E.W. (1957). Interrelations between digestible nutrient and energy content, voluntary dry matter intake, and the overall feeding value of forages. *Journal of Animal Science*, 15, 546–52.

Crespo, D.G. (1982). Legume production. In *Efficient Grassland Farming*, ed. A.J. Corrall, pp. 49–60. Hurley, UK: British Grassland Society Occasional Symposium 14.

Crespo, G. & Cuesta, A. (1974). Effect of different levels of P + K on pangola response to increasing doses of nitrogen fertilization. *Proceedings XII International Grassland Congress*, Vol. 2, pp. 74–82.

Crider, F.J. (1955). *Root Growth Stoppage Resulting from Defoliation of Grass*. Washington: USDA Technical Bulletin 1102.

Crush, J.R. & Campbell, B.D. (1993). Effect of different grass species on nitrogen fixation by white clover under conditions of elevated carbon dioxide and temperature. In *Proceedings XVII International Grassland Congress*, pp. 1130–1.

Crush, J.R., Campbell, B.D. & Evans, J.P.M. (1993). Effect of elevated atmospheric carbon dioxide levels on nodule relative efficiency in white clover. In *Proceedings XVII International Grassland Congress*, pp. 1131–3.

Culvenor, R.A., Davidson, I.A. & Simpson, R.J. (1989a). Regrowth by swards of subterranean clover after defoliation. 1. Growth, non-structural carbohydrate and nitrogen content. *Annals of Botany*, 64, 545–56.

Culvenor, R.A., Davidson, I.A. & Simpson, R.J. (1989b). Regrowth by swards of subterranean clover after defoliation. 2. Carbon exchange in shoot, root and nodule. *Annals of Botany*, 64, 557–67.

Curll, M.L. & Wilkins, R.J. (1982). Frequency and severity of defoliation of grass and clover by sheep at different stocking rates. *Grass and Forage Science*, 37, 291–7.

Curll, M.L., Wilkins, R.J., Snaydon, R.W. & Shanmugalingam, V.S. (1985). The effects of stocking rate and nitrogen fertilizer on a perennial ryegrass – white clover sward. I. Sward and sheep performance. *Grass and Forage Science*, 40, 129–40.

Dalal, R.C. & Mayer, R.J. (1986). Long term trends in fertility of soils under continuous cultivation and cereal cropping in southern Queensland. I. Overall changes in soil properties and trends in winter cereal yields. *Australian Journal of Soil Research*, 24, 265–79.

Danckwerts, J.E. & Gordon, A.J. (1989). Long-term partitioning, storage and remobilisation of ^{14}C assimilated by *Trifolium repens* (cv. Blanca). *Annals of Botany*, 64, 533–44.

Danckwerts, J.E., O'Reagain, P.J. & O'Connor, T.G. (1993). Range management in a changing environment: a southern African perspective. *The Rangeland Journal*, **15**, 133–44.

Date, R.A. & Brockwell, J. (1978). *Rhizobium* strain competition and host interaction for nodulation. In *Plant Relations in Pastures*, ed. J.R. Wilson, pp. 202–16. Melbourne: CSIRO.

Date, R.A., Burt, R.L. & Williams, W.T. (1979). Affinities between various *Stylosanthes* as shown by rhizobial, edaphic and geographic relationships. *Agro-ecosystems*, **5**, 57–67.

Date, R.A. & Halliday, J. (1979). Selecting *Rhizobium* for acid, infertile soils of the tropics. *Nature*, **277**, 62–4.

Date, R.A. & Norris, D.O. (1979). *Rhizobium* screening of *Stylosanthes* species for effectiveness in nitrogen fixation. *Australian Journal of Agricultural Research*, **30**, 85–104.

Datzenko, A. & Ahlgren, G.H. (1951). Effects of cutting treatments on the yield, botanical composition, and chemical constituents of an alfalfa–bromegrass mixture. *Agronomy Journal*, **43**, 15–7.

Davidson, J.L. & Donald, C.M. (1958). The growth of swards of subterranean clover with particular reference to leaf area. *Australian Journal of Agricultural Research*, **9**, 53–72.

Davidson, J.L. & Milthorpe, F.L. (1966a). Leaf growth in *Dactylis glomerata* following defoliation. *Annals of Botany*, **30**, 173–84.

Davidson, J.L. & Milthorpe, F.L. (1966b). The effect of defoliation on the carbon balance in *Dactylis glomerata*. *Annals of Botany*, **30**, 185–98.

Davidson, R.L. (1962). The influence of edaphic factors on the species composition of early stages of the subsere. *Journal of Ecology*, **50**, 401–10.

Davidson, R.L. (1964). Theoretical aspects of nitrogen economy in grazing experiments. *Journal of British Grassland Society*, **19**, 273–80.

Davies, A. (1965). Carbohydrate levels and regrowth in perennial rye-grass. *Journal of Agricultural Science*, **65**, 213–21.

Davies, J.G. (1952). The establishment and maintenance of legumes in the sward. In *Proceedings VI International Grassland Congress*, pp. 433–42.

Davies, J.G. & Sim, A.H. (1931). *The Influence of Frequency of Cutting on the Productivity, Botanical and Chemical Composition, and the Nutritive Value of 'Natural' Pasture*. Melbourne: CSIR Bulletin 48.

Davies, W. (1960). *The Grass Crop*. London: Spon.

Davison, T.M., Cowan, R.T. & Shepherd, R.K. (1985). Milk production from cows grazing on tropical grass pastures. 2. Effects of stocking rate and level of nitrogen fertilizer on milk yield and pasture – milk yield relationship. *Australian Journal of Experimental Agriculture*, **25**, 515–32.

Day, J.M., Neves, M.C.P. & Döbereiner, J. (1975). Nitrogenase activity on the roots of tropical forage grasses. *Soil Biology and Biochemistry*, **7**, 107–12.

de Boer, F. & Bickel, H. ed. (1988). *Livestock Feed Resources and Feed Evaluation in Europe*. Amsterdam: Elsevier.

de Boer, T.A. (1966). Nitrogen effect on the herbage production of grasslands on different sites. In *Proceedings X International Grassland Congress*, pp. 199–204.

de Carvalho, M.M. (1978). A comparative study of the responses of six *Stylosanthes* species to acid soil factors with particular reference to aluminium. PhD thesis, University of Queensland.

de Carvalho, M.M., Edwards, D.G., Andrew, C.S. & Asher, C.J. (1981). Aluminium toxicity, nodulation, and growth of *Stylosanthes* species. *Agronomy Journal*, **73**, 261–5.

de Haan, C. (1994). An overview of the World Bank's involvement in pastoral development. *Pastoral Development Network Paper 36b*, pp. 1–6. London: Overseas Development Institute.

Deregibus, V.A., Sanchez, R.A., Casal, J.J. & Trlica, M.J. (1985). Tillering responses to enrichment of red light beneath the canopy in a humid natural grassland. *Journal of Applied Ecology*, **22**, 199–206.

Detwiler, R.P. & Hall, C.A. (1988). Tropical forests and the global carbon cycle. *Science*, **239**, 42–7.

Dilworth, M.J., Robson, A.D. & Chatel, D.L. (1979). Cobalt and nitrogen fixation in *Lupinus angustifolium* L. II. Nodule formation and function. *New Phytologist*, **83**, 63–79.

Doak, B.W. (1952). Some chemical changes in the nitrogenous constituents of urine when voided on pasture. *Journal of Agricultural Science*, **42**, 162–71.

Donald, C.M. (1946). *Pastures and Pasture Research*. Sydney: University of Sydney.

Dougherty, C.T., Bradley, N.W., Lauriault, L.M., Arias, J.E. & Cornelius, P.L. (1994). Allowance-intake relations of cattle grazing vegetative tall fescue. *Grass and Forage Science*, **47**, 211–19.

Do Valle, C.B., Glienke, C. & Leguizamon, G.O. (1993). Breeding of apomictic *Brachiaria* through interspecific hybridisation. In *Proceedings XVII International Grassland Congress*, pp. 427–8.

Dreyfus, B.L., Diem, H.G., Freire, J., Keya, S.O. & Dommergues, Y.R. (1987). Nitrogen fixation in tropical agriculture and forestry. In *Microbial Technology in the Developing World*, ed. E.J. Da Silva, Y.R. Dommergues, E.J. Nyns, & C. Ratledge, pp. 7–50. Oxford: Oxford University Press.

Dreyfus, B., Garcia, J.L. & Gillis, M. (1988). Characterization of *Azorhizobium caulinodans* gen. nov., sp. nov., a stem-nodulating nitrogen-fixing bacterium isolated from *Sesbania rostrata*. *International Journal of Systematic Bacteriology*, **38**, 89–98.

Drysdale, A.D. (1970). Anhydrous ammonia as a grassland fertilizer. In *Proceedings XI International Grassland Congress*, pp. 424–7.

Duncan, P. & Jarman, P.J. (1993). Conservation of biodiversity in managed rangelands, with special emphasis on the ecological effects of large grazing ungulates, domestic and wild. In *Proceedings XVII International Grassland Congress*, pp. 2077–84.

Duncan, W.G. (1966). A model for simulating photosynthesis and other radiation phenomena in plant communities. In *Proceedings X International Grassland Congress*, pp. 120–5.

Dyksterhuis, E.J. (1949). Condition and management of rangeland based on quantitative ecology. *Journal of Range Management*, **2**, 104–15.

Dyksterhuis, E.J. (1952). Determining the condition and trend of ranges (natural pastures). In *Proceedings VI International Grassland Congress*, pp. 1322–7.

Dyksterhuis, E.J. (1958). Ecological principles in range evaluation. *Botanical Review*, **24**, 253–72.

Ebersohn, J.P., Moir, K.W. & Duncalfe, F. (1985). Inter-relationship between pasture growth and senescence and their effects on live-weight gain of grazing beef cattle. *Journal of Agricultural Science*, **104**, 299–301.

Edmond, D.B. (1966). The influence of animal treading on pasture growth. In *Proceedings X International Grassland Congress*, pp. 453–8.

Edwards, P.J. (1983). Multiple use of grassland resources. (Grasslands to provide natural resource conservation and a quality environment for mankind). In *Proceedings XIV International Grassland Congress*, pp. 64–9.

Edye, L.A., Humphreys, L.R., Henzell, E.F. & Teakle, L.J.H. (1964). Pasture investigations in the Yalleroi district of central Queensland. *University of Queensland Papers of the Department of Agriculture*, 1, 153–72.

Edye, L.A., Williams, W.T., Anning, P., Holm, A.M., Miller, C.P., Page, M.C. & Winter, W.H. (1975). Sward tests of some morphological-agronomic groups of *Stylosanthes* accessions in dry tropical environments. *Australian Journal of Agricultural Research*, 26, 481–96.

Edye, L.A., Williams, W.T. & Winter, W.H. (1978). Seasonal relations between animal gain, pasture production and stocking rate on two tropical grass–legume pastures. *Australian Journal of Agricultural Research*, 29, 103–13.

Ehara, K., Maeno, N. & Yamada, Y. (1966). Physiological and ecological studies on the regrowth of herbage plants. 4. The evidence of utilization of food reserves during the early stage of regrowth in bahiagrass (*Paspalum notatum* Flugge) with $^{14}CO_2$. *Journal of Japanese Society of Grassland Science*, 12, 1–3.

Ellison, L., Croft, A.R. & Bailey, R.W. (1951). *Indicators of Condition and Trend on High Range Watersheds of the Intermountain Range*. USDA Agriculture Handbook 19. Washington: USDA.

Ernst, P., Le Du, Y.L.P. & Carlier, L. (1980). Animal and sward production under rotational and continuous grazing management – a critical appraisal. In *The Role of Nitrogen in Intensive Grassland Production, Proceedings of International Symposium of European Grassland Federation*, pp. 119–26. Wageningen: Pudoc.

Estavillo, J.M., Gonzalez-Murua, C. & Rodriguez, M. (1993). Effects of inorganic nitrogen and cow slurry on yield and nitrogen losses in a natural pasture in the Basque country (Spain). In *Proceedings XVII International Grassland Congress*, pp. 1455–6.

Evans, D.R., Hill, J., Williams, T.A. & Rhodes, I. (1985). Effects of coexistence on the performance of white clover-perennial ryegrass mixtures. *Oecologia*, 66, 536–9.

Evans, D.R., Williams, T.A. & Evans, S.R. (1992). Evaluation of white clover varieties under grazing and their role in farm systems. *Grass and Forage Science*, 47, 342–52.

Evans, G. (1937). Growing pasture types of grasses for seed. In *Report IV International Grassland Congress*, pp. 269–75.

Evans, W.B., Munro, J.M.M. & Scurlock, R.V. (1979). Comparative pasture and animal production from cocksfoot and perennial rye grass varieties under grazing. *Grass and Forage Science*, 34, 64–5.

Eyles, A.G., Cameron, D.G., & J.B. Hacker (eds.) (1985). *Pasture Research in Northern Australia – Its History, Achievements and Future Emphasis*. Brisbane: CSIRO Division of Tropical Crops and Pastures.

Fagan, T.W. & Jones, H.T. (1924). The nutritive value of grasses as shown by their chemical composition. Aberystwyth: Welsh Plant Breeding Station Bulletin Series H No. 3.

Fahey, G.C. Jr., Collins M., Mertens, D.R. & Moser, L.E. (1994). *Forage Quality, Evaluation and Utilization*. Madison, WI: American Society of Agronomy.

Falcinelli, M. (1993). Seed shattering in tall fescue. In *Proceedings XVII International Grassland Congress*, pp. 1666–7.

Fales, S.L., Muller, L.D., O'Sullivan, M., Lanyon, L.E., Hoover, R.J. & Holden, L.A. (1993). Intensive grazing of high-producing Holstein cows: milk production, forage utilization and profit potential at three stocking rates. In *Proceedings XVII International Grassland Congress*, pp. 1309–10.

224   *References*

Fenner, M. (ed.) (1992). *The Ecology of Regeneration in Plant Communities.* Wallingford: CAB International.

Ferguson, J.E., Thomas, D., Andrade, R.P.de, Costa, N.S. & Jutzi, S. (1983). Seed-production potentials of eight tropical pasture species in regions of Latin America. In *Proceedings XIV International Grassland Congress*, pp. 275–8.

Ferguson, J.E., Vera, R. & Toledo, J.M. (1989). *Andropogon gayanus* and *Stylosanthes capitata* in the Colombian llanos – the path from the wild towards adoption. In *Proceedings XVI International Grassland Congress*, pp. 1343–4.

Field, B.F., Chapin III, F.S., Matson, P.A. & Mooney, H.A. (1992). Responses of terrestrial ecosystems to the changing atmosphere: a resource-based approach. *Annual Review of Ecological Systems*, **23**, 201–35.

Figarella, J., Abruna, F. & Vicente-Chandler, J. (1972). Effect of five nitrogen sources applied at four rates to Pangola grass sod under humid tropical conditions. *Journal of Agriculture of University of Puerto Rico*, **56**, 410–16.

Fisher, M.J., Rao, I.M., Ayarza, M.A., Lascano, C.E., Sanz, J.I., Thomas, R.J. & Vera, R.R. (1994). Introduced grass-based pastures in the South American savannas – could they be part of the missing global sink for carbon? *Nature* (London), **371** (6949), 236–8.

Fiske, W.A., D'Souza, G.E., Fletcher, J.J., Phipps, T.T., Bryan, W.B. & Prigge, E.C. (1994). An economic and environmental assessment of alternative forage-resource production systems: a goal-programming approach. *Agricultural Systems*, **45**, 259–70.

Fitchen, J. & Beachy, R.N. (1993). Genetically engineered protection against viruses and fungi. In *Proceedings XVII International Grassland Congress,* pp. 1163–6.

Foin, T.C. (1986). Succession climax and range evaluation in the California prairie ecosystem. In *Rangelands: A Resource Under Siege*, ed. P.J. Joss, P.W. Lynch & O.B. Williams, pp. 5–7. Canberra: Australian Academy of Science.

Foran, B. (1993). A time for change – issues arising from the XVII International Grassland Congress. In *Proceedings XVII International Grassland Congress*, pp. 2251–6.

Foran, B.D., Bastin, G. & Shaw, K.A. (1986). Range assessment and monitoring in arid lands: the use of classification and ordination in range survey. *Journal of Environmental Management*, **22**, 67–84.

Foran, B.D., Tainton, N.M. & Booysen, P. de V. (1978). The development of method for assessing veld condition in three grassveld types in Natal. *Proceedings Grassland Society of Southern Africa*, **13**, 27–33.

Forge, K. (1994). *Grass Check. Grazier Rangeland Assessment for Self-Sustainability.* Brisbane: Department of Primary Industries, Queensland Information Series QI 94005.

Frame, J. & Newbould, P. (1984). Herbage production from grass/white clover swards. In *Forage Legumes*, ed. D.J. Thomson, pp. 15–35. Hurley, UK: British Grassland Society Occasional Symposium 16.

Frankena, H. (1937). Influence of nitrogenous fertilizer on the botanical composition of different types of sward. In *Report IV International Grassland Congress*, pp. 339–44.

Frandsen, K.J. (1952). Theoretical aspects of cross-breeding systems for forage plants. In *Proceedings VI International Grassland Congress*, pp. 306–13.

Fred, E.B., Baldwin, I.L. & McCoy, E. (1932). *Root Nodule Bacteria and Leguminous Plants.* Madison, WI: University of Wisconsin Press.

Fred, E.G. & Graul, E.J. (1916). The effect of soluble nitrogenous salts on nodule formation. *Journal of American Society of Agronomy*, **8**, 316–28.

Frey, K.J. (1992). Plant breeding perspectives for the 1990s. In *Plant Breeding in the 1990's*, ed. H.T. Stalker & J.P. Murphy, pp. 1–16. Wallingford, UK: CAB International.

Freysen, A.H.J. & Woldendorp, J.W. (eds.) (1978). *Structure and Functioning of Plant Populations*. Amsterdam: North-Holland Publishing Co.

Friedel, M.H. (1991). Range condition assessment and the concept of thresholds: a viewpoint. *Journal of Range Management*, **44**, 422–6.

Fujita, H. & Humphreys, L.R. (1992). Variation in seasonal stocking rate and the dynamics of *Lotononis bainesii* in *Digitaria decumbens* pastures. *Journal of Agricultural Science*, **118**, 47–53.

Gammon, D.M. (1978). A review of experiments comparing systems of grazing management on natural pastures. *Proceedings of Grassland Society of Southern Africa*, **13**, 75–82.

Gammon, D.M. (1984). An appraisal of short duration grazing as a method of veld management. *Zimbabwe Agricultural Journal*, **81**, 59–64.

Gammon, D.M. & Roberts, B.R. (1978). Patterns of defoliation during continuous and rotational grazing of the Matopos sandveld of Rhodesia. 3. Frequency of defoliation. *Rhodesian Journal of Agricultural Research*, **16**, 147–64.

Gammon, D.M. & Twiddy, D.R. (1990). Patterns of defoliation in four- and eight-paddock grazing systems. *Journal of Grassland Society of Southern Africa*, **7**, 29–35.

Gardener, C.J. (1982). Population dynamics and stability of *Stylosanthes hamata* cv. Verano in grazed pastures. *Australian Journal of Agricultural Research*, **33**, 63–74.

Garnsworthy, P.C. (ed.) (1988). *Nutrition and Lactation in the Dairy Cow*. London: Butterworths.

Gartner, J.A. (1969). Effect of fertilizer nitrogen on a dense sward of Kikuyu, paspalum and carpet grass. 1. Botanical composition, growth and nitrogen uptake. *Queensland Journal of Agricultural and Animal Sciences*, **26**, 21–33.

Gates, D.M. (1993). *Climate Change and its Biological Consequences*. Sunderland, MA: Sinauer.

Gervais, P. (1960). Effects of cutting treatments on ladino clover grown alone and in mixture with grasses. I. Productivity and botanical composition of forage. *Canadian Journal of Plant Science*, **40**, 317–27.

Gibson, A.H. (1966). The carbohydrate requirements for symbiotic nitrogen fixation: a 'whole-plant' growth analysis approach. *Australian Journal of Biological Sciences*, **19**, 499–515.

Gibson, A.H. & Nutman, P.S. (1960). Studies on the physiology of nodule formation. VII. A reappraisal of the effect of pre-planting. *Annals of Botany*, **24**, 420–33.

Gibson, T. (1987). Northeast Thailand. A ley farming system using dairy cattle in the infertile uplands. *World Animal Review*, **61**, 36–43.

Gibson, T.A. & Humphreys, L.R. (1973). The influence of nitrogen nutrition of *Desmodium uncinatum* on seed production. *Australian Journal of Agricultural Research*, **24**, 667–76.

Gifford, R.M. (1992). Interaction of carbon dioxide with growth-limiting environmental factors in vegetation productivity: implications for the global carbon cycle. *Advances in Bioclimatology*, **1**, 25–58.

Gifford, R.M. (1993). Climate change – predicting the effects. In *Proceedings XVII Grassland Congress*, pp. 1155–6.

226   *References*

Gifford, R.M. (1994). The global carbon cycle: a viewpoint on the missing sink. *Australian Journal of Plant Physiology*, 21, 1–15.
Gifford, R.M. & Marshall, C. (1973). Photosynthesis and assimilate distribution in *Lolium multiflorum* Lam. following differential tiller defoliation. *Australian Journal of Biological Sciences*, 26, 517–26.
Gilbert, M.A. & Clarkson, N.M. (1993). Efficient nitrogen fertiliser strategies for tropical beef and dairy production using summer rainfall and soil analyses. In *Proceedings XVII International Grassland Congress*, pp. 1552–3.
Gillard, P. (1966). Responses to grazing intensity on the Transvaal highveld. *Experimental Agriculture*, 2, 217–24.
Gillard, P. (1969). The effect of stocking rate on botanical composition and soils in natural grassland in South Africa. *Journal of Applied Ecology*, 6, 489–97.
Gillen, R.L., McCollum, F.T., Hodges, M.E., Bremmer, J.E. & Tate, K.W. (1991). Plant community responses to short duration grazing in tallgrass prairie. *Journal of Range Management*, 44, 124–8.
Giller, K.E. & Wilson, K.J. (1991). *Nitrogen Fixation in Tropical Cropping Systems*. Wallingford, UK: CAB International.
Giöbel, G. (1937). Experiments on the use of nitrogen on Swedish pastures. In *Report IV International Grassland Congress*, pp. 330–8.
Godwin, I.D., Cameron, D.F. & Gordon, G.H. (1990). Variation among somaclonal progenies from three species of *Stylosanthes*. *Australian Journal of Agricultural Research*, 41, 645–56.
Good, R. (1964). *The Geography of the Flowering Plants*, 4th edn. London: Longman.
Gordon, I.J. & Lascano, C. (1993). Foraging strategies of ruminant livestock on intensively managed grasslands: potential and constraints. In *Proceedings XVII International Grassland Congress*, pp. 681–9.
Goudriaan, J. (1990). Atmospheric CO_2, global carbon fluxes and the biosphere. In *Theoretical Production Ecology: Reflections and Prospects*, ed. R. Rabbinge, J. Goudriaan, H. van Keulen, F.W.T. Penning de Vries & H.H. van Laar, pp. 17–14. Simulation monographs 34. Wageningen: PUDOC.
Goudriaan, J. (1992). Biosphere structure, carbon sequestering potential and the atmospheric ^{14}C carbon record. *Journal of Experimental Botany*, 43, 1111–19.
Graber, L.F., Nelson, N.T., Leukel, W.A. & Albert, N.B. (1927). Organic food reserves in relation to the growth of alfalfa and other perennial herbaceous plants. *Agricultural Experiment Station of University of Wisconsin Research Bulletin 8*.
Grace, J.B. (1991). A clarification of the debate between Grime and Tilman. *Functional Ecology*, 5, 583–7.
Graetz, R.D. (1986). A comparative study of sheep grazing a semi-arid saltbush pasture in two condition classes. *Australian Rangeland Journal*, 8, 46–56.
Graham, T.G. (1951). Tropical pasture investigations. *Queensland Agricultural Journal*, 73, 311–26.
Graham, T.W.G., Webb, A.A. & Waring, S.A. (1981). Soil nitrogen status and pasture productivity after clearing of brigalow (*Acacia harpophylla*). *Australian Journal of Experimental Agriculture and Animal Husbandry*, 21, 109–18.
Grant, S.A. Barthram, G.T., Torvell, L., King, J. & Smith, H.K. (1983). Sward management, lamina turnover and tiller population density in continuously stocked *Lolium perenne*-dominated swards. *Grass and Forage Science*, 38, 333–44.

Green, B.H. (1990). Agricultural intensification and the loss of habitat, species and amenity in British grasslands: a review of historical change and assessment of future prospects. *Grass and Forage Science*, **45**, 365–72.

Green, J.O. & Cowling, D.W. (1960). The nitrogen nutrition of grassland. In *Proceedings VIII International Grassland Congress*, pp. 126–9.

Greenhalgh, J.F.D., Reid, G.W., Aitken, J.N. & Florence, E. (1966). The effects of grazing intensity on herbage consumption and animal production. I. Short-term effects in strip-grazed cows. *Journal of Agricultural Science*, **67**, 13–23.

Gregory, F.G. (1938). The constancy of mean net assimilation rate and its ecological importance. *Annals of Botany*, **2**, 811–

Griffith, G. ap & Walters, R.J.K. (1966). The sodium and potassium content of some grass genera, species and varieties. *Journal of Agricultural Science*, **67**, 81–9.

Griffiths, D.J. (1956). Standards of spatial isolation for seed production of herbage crops. *Herbage Abstracts*, **26**, 205–12.

Griffiths, D.J., Lewis, J. & Bean, E.W. (1980). Problems of breeding for seed production in grasses. In *Seed Production*, ed. P.D. Hebblethwaite, pp. 37–50. London: Butterworths.

Grime, J.P. (1977). Evidence for the existence of three primary strategies in plants and its relevance to ecological and evolutionary theory. *American Naturalist*, **111**, 1169–94.

Grime, J.P. (1979). *Plant Strategies and Vegetation Processes*. Chichester: Wiley.

Grime, J.P., Hodgson, J.G. & Hunt, R. (1988). *Comparative Plant Ecology: A Functional Approach to Common British Species*. London: Unwin, Hyman.

Grunow, J.O., Pienaar, A.J. & Breytenbach, C. (1970). Long term nitrogen application to veld in South Africa. *Proceedings Grassland Society of Southern Africa*, **5**, 75–90.

Guerrero, J.W., Conrad, B.E., Holt, E.C. & Wu, H. (1984). Prediction of animal performance on bermuda grass pasture from available forage. *Agronomy Journal*, **76**, 577–80.

Gutteridge, R.C. (1992). Evaluation of the leaf of a range of tree legumes as a source of nitrogen for crop growth. *Experimental Agriculture*, **28**, 195–202.

Gutteridge, R.C. & Shelton, H.M. (1994). *Forage Tree Legumes in Tropical Agriculture*. Wallingford, UK: CAB International.

Hacker, J.B. (1982). Selecting and breeding better quality grasses. In *Nutritional Limits to Animal Production from Pastures*, ed. J.B. Hacker, pp. 305–26. Farnham Royal, UK: CAB.

Hacker, J.B., Forde, B.J. & Gow, J.M. (1974). Simulated frosting of tropical grasses. *Australian Journal of Agricultural Research*, **25**, 45–57.

Hacker, J.B. & Ternouth, J.H. (eds.) (1987). *The Nutrition of Herbivores*. Sydney: Academic Press.

Hacker, R. (1993). A brief evaluation of time control grazing. In *Proceedings Eighth Annual Conference of Grassland Society of NSW*, pp. 82–9.

Hadley, M. (1993). Grasslands for sustainable ecosystems. In *Proceedings XVII International Grassland Congress*, pp. 21–7.

Hamilton, W.T. & Seifres, C.J. (1982). Prescribed burning during winter for maintenance of buffelgrass. *Journal of Range Management*, **35**, 9–12.

Hanley, J.A. (1937). The need for lime and phosphate in grassland improvement. In *Report IV International Grassland Congress*, pp. 288–97.

Hanson, J. & Lazier, J.R. (1989). Forage germplasm at the International Livestock Centre for Africa (ILCA): an essential resource for evaluation and selection. In *Proceedings XVI International Grassland Congress*, pp. 265–6.

Harlan, J.R. (1958). Generalised curves for gain per head and gain per acre in rates of grazing studies. *Journal of Range Management*, 11, 140–7.

Harlan, J.R. (1983a). The scope for collection and improvement of forage plants. In *Genetic Resources of Forage Plants*, ed. J.G. McIvor & R.A. Bray, pp. 3–14. Melbourne: CSIRO.

Harlan, J.R. (1983b). Use of genetic resources for improvement of forage species. In *Proceedings XIV International Grassland Congress*, pp. 29–34.

Harper, J.L. (1977). *Population Biology of Plants*. London: Academic Press.

Harper, J.L. (1978). Plant relations in pastures. In *Plant Relations in Pastures*, ed. J.R. Wilson, pp. 3–14. Melbourne: CSIRO.

Harrington, G.N. & Pratchett, D. (1974a). Stocking rate trials in Ankole, Uganda. 1. Weight gain of Ankole steers at intermediate and heavy stocking rates under different managements. *Journal of Agricultural Science*, 82, 497–506.

Harrington, G.N. & Pratchett, D. (1974b). Stocking rate trials in Ankole, Uganda. II. Botanical analysis and oesophageal fistula sampling of pastures grazed at different stocking rates. *Journal of Agricultural Science*, 82, 507–16.

Hart, R.H., Bissio, J., Samuel, M.J. & Waggoner, Jr J.W. (1991). Grazing systems, pasture size, and cattle grazing behavior, distribution and gains. *Journal of Range Management*, 46, 81–7.

Hart, R.H. & Hanson, J.D. (1993). Managing for economic and ecological stability of range and range-improved grassland systems with the SPUR2 model and the STEERISKIER spreadsheet. In *Proceedings XVII International Grassland Congress*, 1593–8.

Hartley, W. (1950). The global distribution of tribes of the Gramineae in relation to historical and environmental factors. *Australian Journal of Agricultural Research*, 1, 355–73.

Hartley, W. (1954). The agrostological index: a phytogeographical approach to the problems of pasture plant introduction. *Australian Journal of Botany*, 2, 1–21.

Hartley, W. (1963). The phytogeographical basis of pasture plant introduction. *Genetica agraria*, 17, 135–60.

Hartley, W. & Slater, C. (1960). Studies on the origin, evolution, and distribution of the Gramineae. III. The tribes of the subfamily Eragrostoideae. *Australian Journal of Botany*, 8, 256–76.

Hartwig, U., Boller, B.C., Baur-Höch, B. & Nösberger, J. (1990). The influence of carbohydrate reserves on the response of nodulated clover to defoliation. *Annals of Botany*, 65, 97–105.

Hartwig, U., Boller, B.C. & Nösberger, J. (1987). Oxygen supply limits nitrogenase activity of clover nodules after defoliation. *Annals of Botany*, 59, 285–91.

Hatch, G.P. & Tainton, W.M. (1993). A bioeconomic stocking rate model for the semi-arid savanna of Natal, South Africa. In *Proceedings XVII International Grassland Congress*, pp. 59–60.

Hatch, M.D. & Slack, C.R. (1966). Photosynthesis by sugar cane leaves. A new carboxylation reaction and the pathway of sugar formation. *Biochemical Journal*, 101, 103.

Hatch, S.E., Jarvis, S.C. & Reynolds, S.E. (1991). An assessment of the contribution of net mineralization to N cycling in grass swards using a field incubation method. *Plant and Soil*, 138, 28–32.

Hayward, M.D. & Breese, E.L. (1968). The genetic organisation of natural populations of *Lolium perenne* L. *Heredity*, 23, 357–68.

Heady, H.F. (1958). Vegetational changes in the California annual type. *Ecology*, 39, 402–15.

Heady, H.F. (1960). Range management in the semi-arid tropics of East Africa according to principles developed in temperate climates. In *Proceedings VIII International Grassland Congress*, pp 223–6.

Heady, H.F. (1975). *Rangeland Management*. New York: McGraw Hill.

Heddle, R.G. & Ogg, W.G. (1937). Soil nutrients in relation to pasture maintenance and improvement. In *Report IV International Grassland Congress*, pp. 298–302.

Hegarty, M.P. (1982). Deleterious factors in forages affecting animal production. In *Nutritional Limits to Animal Production from Pastures*, ed. J.B. Hacker, pp. 133–50. Farnham Royal, UK: Commonwealth Agricultural Bureaux.

Heitschmidt, R.K., Frasure, J.R., Price, D.Z. & Rittenhouse, L.R. (1982). Short duration grazing at the Texas Experimental Ranch: weight gains of growing heifers. *Journal of Range Management*, **35**, 375–9.

Hendershot, K.L. & Volenc, J.J. (1989). Shoot growth, dark respiration, and nonstructural carbohydrates of contrasting alfalfa genotypes. *Crop Science*, **29**, 1271–5.

Hengeveld, H. & Kertland, P. (1995). An assessment of new developments relevant to the science of climate change. *Climate Change Newsletter*, **7**(3), 1–24.

Hennessy, D.W. (1980). Protein nutrition of ruminants in tropical areas of Australia. *Tropical Grasslands*, **14**, 260–5.

Henzell, E.F. (1970). Problems in comparing the nitrogen economies of legume-based and nitrogen-fertilized pasture systems. In *Proceedings XI International Grassland Congress*, pp. A112–20.

Henzell, E.F. (1971). Recovery of nitrogen from four fertilizers applied to Rhodes grass in small plots. *Australian Journal of Experimental Agriculture and Animal Husbandry*, **11**, 420–30.

Henzell, E.F. (1988). The role of biological nitrogen fixation research in solving problems in tropical agriculture. *Plant and Soil*, **108**, 15–21.

Henzell, E.F., Fergus, I.F. & Martin, A.E. (1966). Accumulation of soil nitrogen and carbon under a *Desmodium uncinatum* pasture. *Australian Journal of Experimental Agriculture and Animal Husbandry*, **6**, 157–60.

Henzell, E.F. & Oxenham, D.J. (1964). Seasonal changes in the nitrogen content of three warm-climate pasture grasses. *Australian Journal of Experimental Agriculture and Animal Husbandry*, **4**, 336–44.

Henzell, E.F., Peake, D.C.I., Mannetje, L.'t. & Stirk, G.B. (1975). Nitrogen response of pasture grasses on duplex soils formed from granite in southern Queensland. *Australian Journal of Experimental Agriculture and Animal Husbandry*, **15**, 498–507.

Herridge, D.F. & Bergersen, F.J. (1988). Symbiotic nitrogen fixation. In *Advances in Nitrogen Cycling in Agricultural Ecosystems*, ed. J.R. Wilson, pp. 46–65. Wallingford, UK: CAB International.

Herridge, D.F., Bergersen, F.J. & Peoples, M.B. (1990). Measurement of nitrogen fixation by soybean in the field using the ureide and natural ^{15}N abundance methods. *Plant Physiology*, **93**, 708–16.

Hesky, L.E., Janovsky, J., Gyulai, G., Kiss, E. & Hangyel, L.T. (1993). Application of the somaclone method in fertility restoration of the partially sterile hybrid of *Agropyron repens* × *Bromus inermis*: callus initiation and plant regeneration. In *Proceedings XVII International Grassland Congress*, pp. 1048–9.

Hides, D.H. & Desroches, R. (1989). The role of seeds in forage production factors limiting optimal utilization. In *Proceedings XVI International Grassland Congress*, pp. 1777–84.

Hildreth, R.J. & Riewe, M.E. (1963). Grazing production curves. II. Determining the economic optimum stocking rate. *Agronomy Journal*, **55**, 370–2.

Hill, M.J. & Loch, D.S. (1993). Achieving potential herbage seed yields in tropical regions. In *Proceedings XVII International Grassland Congress*, pp. 1629–35.

Hirata, M., Sugimoto, Y. & Ueno, M. (1986). Energy and matter flows in bahiagrass pasture. II. Net primary production and efficiency for solar energy utilisation. *Journal of Japanese Society of Grassland Science*, **31**, 387–96.

Hoden, A., Peyraud, J.L., Muller, A., Delaby, L., Faverdin, P., Peccate, J.R. & Fargetton, M. (1991). Simplified rotational grazing management of dairy cows: effects of rates of stocking and concentrate. *Journal of Agricultural Science*, **116**, 417–28.

Hodges, T.K., Rathore, K.S. & Peng, Y. (1993). Advances in genetic transformation of plants. In *Proceedings XVII International Grassland Congress*, pp. 1013–23.

Hodgkinson, K.C. (1970). Physiological aspects of the regeneration of lucerne. In *Proceedings XI International Grassland Congress*, pp. 559–61.

Hodgkinson, K.C. (1974). Influence of partial defoliation on photosynthesis, photorespiration and transpiration by lucerne leaves of different ages. *Australian Journal of Plant Physiology*, **1**, 561–78.

Hodgkinson, K.C., Ludlow, M.M., Mott, J.J. & Baruch, Z. (1989). Comparative responses of the savanna grasses *Cenchrus ciliaris* and *Themeda triandra* to defoliation. *Oecologia*, **79**, 45–52.

Hodgson, J. (1981). Variation in the surface characteristics of the sward and the short-term rate of herbage intake by calves and lambs. *Grass and Forage Science*, **36**, 49–57.

Hodgson, J. (1985). The significance of sward characteristics in the management of temperate sown pastures. In *Proceedings XV International Grassland Congress*, pp. 63–7.

Hodgson, J., Mackie, C.K. & Parker, J.W.G. (1986). Sward surface heights for efficient grazing. *Grass Farmer*, **24**, 5–10.

Hoffman, M.T. (1988). The rationale for karoo grazing systems: criticisms and research implications. *South African Journal of Science*, **84**, 556–9.

Holling, C.S. (1973). Resilience and stability of ecological systems. *Annual Review of Ecology and Systematics*, **4**, 1–23.

Holmes, J.H. & Day, P. (1995). Identity, lifestyle and survival: value orientations of South Australian pastoralists. *The Rangeland Journal*, **17**, 193–212.

Holt, J.E. & Schoorl, D. (1990). The application of open and closed systems theory to change in agricultural institutions. *Agricultural Systems*, **34**, 123–32.

Hoogendorn, C.J., Holmes, C.W. & Chu, A.C.P. (1992). Some effects of herbage composition, as influenced by previous grazing management, on milk production by cows grazing on rye grass/white clover pastures. 2. Milk production in late spring/summer: effects of grazing intensity during the preceding spring period. *Grass and Forage Science*, **47**, 316–25.

Hopkins, A., Davies, A. & Doyle, C. (1994). *Clovers and Other Grazed Legumes in UK Pasture Land*. IGER Technical Review No. 1. Aberystwyth: IGER.

Hopkins, A., Wainwright, J., Murray, P.J., Bowling, P.J. & Webb, M. (1988). 1986 survey of upland grassland in England and Wales: changes in age structure and botanical composition since 1970–72 in relation to grassland management and physical features. *Grass and Forage Science*, **43**, 185–98.

Hopkinson, J.M. & Reid, R. (1979). Significance of climate in tropical pasture/legume seed production. In *Pasture Production in Acid Soils of the Tropics*, ed. P.A. Sánchez & L.E. Tergas, pp. 343–60. Cali, Colombia: CIAT.

Houghton, J.T., Jenkins, G.J. & Ephraums, J.J. (eds.) (1990). *Climate Change, the IPCC Scientific Assessment*. Cambridge: Cambridge University Press.

Houghton, J.T., Callander, B.A. & Varney, S.K. (eds.) (1992). *Climate Change 1992: the Supplementary Report to the IPCC Scientific Assessment. Working Group 1. Bracknell*. Cambridge: Cambridge University Press.

Howden, S.M., White, D.H., McKeon, G.M., Scanlan, J.C. & Carter, J.O. (1994). Methods for exploring management options to reduce greenhouse gas emissions from tropical grazing systems. *Climate Change*, **27**, 49–70.

Hull, J.L., Meyer, J.H. & Kromann, R. (1961). Influence of stocking rate on animal and forage production from irrigated pasture. *Journal of Animal Science*, **20**, 46–52.

Humphrey, R.R. (1949). Field comments on the range condition method of forage survey. *Journal of Range Management*, **2**, 1–10.

Humphrey, R.R. (1962). *Range Ecology*. New York: Ronald Press.

Humphreys, L.R. (1966a). Sub-tropical grass growth. II. Effects of variation in leaf area index in the field. *Queensland Journal of Agricultural and Animal Sciences*, **23**, 337–58.

Humphreys, L.R. (1966b). Sub-tropical grass growth. III. Effects of stage of defoliation and inflorescence removal. *Queensland Journal of Agricultural and Animal Sciences*, **23**, 499–531.

Humphreys, L.R. (1967). Townsville lucerne – history and prospect. *Journal of Australian Institute of Agricultural Science*, **33**, 3–13.

Humphreys, L.R. (1981). *Environmental Adaptation of Tropical Pasture Plants*. London: MacMillan.

Humphreys, L.R. (1989). Future directions in grassland science and its applications. In *Proceedings XVI International Grassland Congress*, **3**, 1705–10.

Humphreys, L.R. (1991). *Tropical Pasture Utilisation*. Cambridge: Cambridge University Press.

Humphreys, L.R. (1994). *Tropical Forages: Their Role in Sustainable Agriculture*. Harlow, UK: Longman. New York: John Wiley.

Humphreys, L.R. & Riveros, F. (1986). *Tropical Pasture Seed Production*. Rome: FAO Plant Production and Protection Paper 8.

Humphreys, L.R. & Robinson, A.R. (1966). Sub-tropical grass growth. 1. Relationship between carbohydrate accumulation and leaf area in growth. *Queensland Journal of Agricultural and Animal Sciences*, **23**, 211–59.

Humphreys, M.O., Humphreys, M.W. & Thomas, H. (1993). Breeding grasses for adaptation to environmental problems. In *Proceedings XVII International Grassland Congress*, pp. 441–2.

Humphries, E.C. (1963). Dependence of net assimilation rate on root growth of isolated leaves. *Annals of Botany*, **27**, 125–

Hussey, M.A., Bashaw, E.C., Highnight, K.W., Wipff, J. & Hatch, S.L. (1993). Fertilization of unreduced female gametes: a technique for genetic enhancement within the *Cenchrus–Pennisetum* agamic complex. In *Proceedings XVII International Grassland Congress*, pp. 404–5.

Hutton, E.M. & Beall, L.B. (1977). Breeding of *Macroptilium atropurpureum*. *Tropical Grasslands*, **11**, 15–31.

Hutton, E.M. & Bonner, I.A. (1960). Dry matter and protein yields in four strains of *Leucaena glauca* Benth. *Journal of Agricultural Institute of Agricultural Science*, **26**, 276–7.

IBPGR (1992). *Descriptors for Annual* Medicago. Rome: IBPGR.

Idso, S.B. (1992). Shrubland expansion in the American southwest. *Climatic Change*, 22, 85–6.

IGC (1937). *Report IV International Grassland Congress*, 15–17 July 1937, Aberystwyth, UK.

IGC (1952). *Report VI International Grassland Congress*, 17–23 August 1952, State College, Pennsylvania, USA.

IGC (1966). *Report X International Grassland Congress*, 7–16 July 1966, Helsinki, Finland: Finnish Grassland Association.

IGC (1981). *XIV International Grassland Congress. Summaries of Papers*, ed. J.A. Smith, June 14–24, 1981. Lexington, Kentucky: University of Kentucky College of Agriculture.

IGC (1983). *Proceedings XIV International Grassland Congress*, 15–24 June, 1981, Lexington, Kentucky, ed. J.A. Smith & V.W. Hays, Boulder, Colorado: Westview Press.

IGC (1993). *Proceedings XVII International Grassland Congress*, 8–21 February 1993, Palmerston North, Hamilton, Lincoln, New Zealand and Rockhampton, Australia. Palmerston North, NZ: New Zealand Grassland Association.

Iglovikov, V.G., Kulakov, V.A. & Blagoveschchensky, G.V. (1985). Effect of anhydrous ammonia (NH_3) and nitrifying inhibitor on productivity of grass pastures. In *Proceedings XV International Grassland Congress*, pp. 481–2.

Illius, A.W., Lowman, B.G. & Hunter, E.A. (1987). Control of sward conditions and apparent utilization of energy in the buffer grazing system. *Grass and Forage Science*, 42, 283–95.

Institute for Grassland and Animal Production (1988). *Report 1987. Volume 2. The Welsh Plant Breeding Station.* Hurley, UK: AFRC Institute for Grassland and Animal Production.

Institute for Grassland and Animal Production (1990). *Report 1989*, p. 3. Aberystwyth, UK: AFRC Institute of Grassland and Environmental Research.

Institute of Grassland and Environmental Research (1992). *1991 Report.* Aberystwyth, UK: Institute of Grassland and Environmental Research.

Institute of Grassland and Environmental Research (1993). *1992 Report.* Aberystwyth, UK: AFRC Institute of Grassland and Environmental Research.

Institute of Grassland and Environmental Research (1994). *1993 Annual Report.* Aberystwyth, UK: IGER.

Irwin, J.A.C. & Cameron, D.F. (1978). Two diseases of *Stylosanthes* spp. caused by *Colletotrichum gloeosporioides* in Australia, and pathogenic specialization within one of the causal organisms. *Australian Journal of Agricultural Research*, 29, 305–17.

Ison, R.L. (1993). Soft systems – a non-computer view of decision support. In *Decision Support Systems for the Management of Grazing Lands*, ed. J.W. Stuth & B.G. Lyons, pp. 83–119. Carnforth, UK: UNESCO Man and the Biosphere Book 11, Parthenon Publishing.

Ison, R.L. & Humphreys, L.R. (1984). Day and night temperature control of floral induction in *Stylosanthes guianensis* cv. Schofield. *Annals of Botany*, 53, 207–11.

Ivory, D.A. & Whiteman, P.C. (1978). Effect of temperature on growth of five subtropical grasses. 1. Effect of day and night temperature on growth and morphological development. *Australian Journal of Plant Physiology*, 5, 131–48.

Izak, A-M.N., Anaman, K. & Jones, R.J. (1990). Biological and economic optima in a tropical grazing ecosystem in Australia. *Agriculture, Ecosystems and the Environment*, **30**, 265–79.

Jarvis, S.C. (1993). Nitrogen cycling and losses from dairy farms. *Soil Use and Management*, **9**(3), 99–105.

Jarvis, S.C., Barraclough, D., Williams, J. & Rook, A.J. (1991). Patterns of denitrification loss from grazed grassland: effects of N fertilizer inputs at different sites. *Plant and Soil*, **131**, 77–88.

Jarvis, S.C., Hatch, D.J. & Dollard, G.J. (1993). Greenhouse gas exchanges with temperate grassland systems. In *Proceedings XVII International Grassland Congress*, pp. 1197–8.

Jarvis, S.C., Hatch, D.J., Orr, R.J. & Reynolds, S.E. (1991). Micro-meteorological studies of ammonia emission from sheep grazed swards. *Journal of Agricultural Science*, **117**, 101–9.

Jarvis, S.C., Hatch, D.J., Pain, B.F. & Klarenbeek, J.V. (1994). Denitrification and the evolution of nitrous oxide after the application of cattle slurry to a peat soil. *Plant and Soil*, **166**, 231–41.

Jarvis, S.C. & Robson, A.D. (1983). The effects of nitrogen nutrition of plants on the development of acidity in Western Australian soils. 1. Effects with subterranean clover grown under leaching conditions. *Australian Journal of Agricultural Research*, **34**, 341–53.

Jarvis, S.C., Scholefield, D. & Pain, B. (1995). Nitrogen cycling in grazing systems. In *Nitrogen Fertilization in the Environment*, ed. P.E. Bacon, pp. 381–419. New York: Marcel Dekker.

Jenkinson, D.S., Adams, D.E. & Wild, A. (1992). Model estimates of CO_2 emissions from soil in response to global warming. *Nature*, **351**, 304–6.

Jenkinson, D.S., Fox, R.H. & Rayner, J.H. (1985). Interaction between fertilizer nitrogen and soil nitrogen – the so-called 'priming' effect. *Journal of Soil Science*, **36**, 425–44.

Jewiss, O.R. (1972). Tillering in grasses – its significance and control. *Journal of British Grassland Society*, **27**, 65–82.

Jewiss, O.R. & Woledge, J. (1967). The effect of age on the rate of apparent photosynthesis in leaves of tall fescue (*Festuca arundinacea* Schreb.). *Annals of Botany*, **31**, 661–71.

Jiggins, J. (1993). From technology transfer to resource management. In *Proceedings XVII International Grassland Congress*, pp. 615–22.

Johnson, D.A., Asoy, K.H., Tieszen, L.L., Ehleringer, J.R. & Jefferson, P.G. (1990). Carbon isotope discrimination: potential in screening cool-season grasses for water-limited environments. *Crop Science*, **30**, 338–43.

Johnson, L.A.Y. & Leatch, G. (1975). Effect of different tick control techniques on tick populations and cattle productivity. In *Annual Report 1975*, pp. 60–1. CSIRO Division of Animal Health.

Johnson, R.W. (1964). *Ecology and Control of Brigalow in Queensland*. Brisbane: QDPI.

Jones, C.A. & Carabaly, A. (1981). Some characteristics of the regrowth of 12 tropical grasses. *Tropical Agriculture*, **58**, 37–44.

Jones, J.G.W. & Baker, R.D. (1966). An integrative approach to research in grassland production and utilization. In *Proceedings X International Grassland Congress*, pp. 510–14.

Jones, Ll. Iorweth (1937). The comparative value of pastures as measured by the grazing animal. In *Report IV International Grassland Congress*, pp. 386–94.

Jones, M. (1937). The improvement of grassland by its proper management. In *Report IV International Grassland Congress*, pp. 470–3.

Jones, M.B., Long, S.P. & Roberts, M.J. (1992). Synthesis and conclusions. In *Primary Productivity of Grass Ecosystems of the Tropics and Sub-tropics*, ed. S.P. Long, M.B. Jones & M.J. Roberts, pp. 212–55. London: Chapman & Hall.

Jones, M.Ll. & Humphreys, M.O. (1993). Progress in breeding interspecific hybrid ryegrasses. *Grass and Forage Science*, **48**, 18–25.

Jones, R.J. (1982). The effect of rust (*Uromyces appendiculatus*) on the yield and digestibility of *Macroptilium atropurpureum* cv. Siratro. *Tropical Grasslands*, **16**, 130–5.

Jones, R.J. & Jones, R.M. (1989). Liveweight gain from rotationally and continuously grazed pastures of *Narak setaria* and Samford rhodes grass fertilized with nitrogen in south east Queensland. *Tropical Grasslands*, **23**, 135–42.

Jones, R.J. & Megarrity, R.G. (1986). Successful transfer of DHP-degrading bacteria from Hawaiian goats to Australian ruminants. *Australian Veterinary Journal*, **63**, 259–62.

Jones, R.J. & Sandland, R.L. (1974). The relation between animal gain and stocking rate. Derivation of the relation from the results of grazing trials. *Journal of Agricultural Science*, **83**, 335–42.

Jones, R.M. (1984). White clover (*Trifolium repens*) in subtropical south-east Queensland. III. Increasing clover and animal production by use of lime and flexible stocking rates. *Tropical Grasslands*, **18**, 186–93.

Jones, R.M. (1988). The effect of stocking rate on the population dynamics of Siratro in Siratro (*Macroptilium atropurpureum*) – Setaria (*Setaria sphacelata*) pastures in south-east Queensland. 3. Effects of spelling on restoration of Siratro in overgrazed pastures. *Tropical Grasslands*, **22**, 5–11.

Jones, R.M. (1992). Resting from grazing to reverse changes in sown pasture composition: application of the 'state-and-transition' model. *Tropical Grasslands*, **26**, 97–9.

Jones, R.M. & Ratcliff, D. (1983). Patchy grazing and its relation to deposition of cattle dung pats in pastures in coastal sub-tropical Queensland. *Journal of the Australian Institute of Agricultural Science*, **49**, 109–11.

Jordan, C.F. (ed.) (1981). *Tropical Ecology*. Stroudsberg: Hutchinson Ross.

Joyce, L.A. (1993). The life cycle of the range condition concept. *Journal of Range Management*, **46**, 132–8.

Jung, H.G., Buxton, D.R., Hatfield, R.D. & Ralph, J. (ed.) (1993). *Forage Cell Wall Structure and Digestibility*. Madison, WI: American Society of Agronomy.

Kaltofen, H., Kreil, W., Kasdorff, K. & Leistner, J. (1966). The effect of heavy doses of nitrogen applied to pasture in spring compared with split applications given during the vegetative period. In *Proceedings X International Grassland Congress*, pp. 231–4.

Kamnalrut, A. & Evenson, J.P. (1992). Monsoon grassland in Thailand. In *Primary Productivity of Grass Ecosystems of the Tropics and Sub-Tropics*, ed. S.P. Long, M.B. Jones & M.J. Roberts, pp. 100–26. London: Chapman & Hall.

Keller-Grein, G., Amézquita, M.C., Lema, G. & Franco, L.H. (1993). Multilocational testing of grasses and legumes in the humid tropics of South America. In *Proceedings XVII International Grassland Congress*, pp. 217–19.

Keswani, C.L. & Mreta, R.A.D. (1980). Effect of intercropping on the severity of powdery mildew on green-gram. *Proceedings of the Second Symposium on Intercropping in Semi-arid Areas, 4–7 August 1980*, ed. C.L. Keswani & B.J. Ndunguru, pp. 110–4. Ottawa, Canada: IDRC 186e.

Khan, M.R.I., Ceriotti, A., Tabe, L., Aryan, A., McNabb, W., Moore, A., Craig, S., Spencer, D. & Higgins, T.J.V. (1996). Accumulation of a sulfur-rich seed albumin from sunflower in the leaves of transgenic subterranean clover (*Trifolium subterraneum* L.). *Transgenic Research*, 5, 79–85.

Klebesadel, L.J. & Helm, D. (1986). Food reserve storage, low temperature injury, winter survival, and forage yields of timothy in subarctic Alaska as related to latitude-of-origin. *Crop Science*, 26, 325–34.

Knott, J.C., Hodgson, R.E. & Ellington, E.V. (1934). *Washington Agricultural Experiment Station Bulletin 295*. Washington, DC: USDA.

Knox, J.P. & Wareing, P.E. (1984). Apical dominance in *Phaseolus vulgaris* L. The possible roles of abscisic and indole-3–acetic acid. *Journal of Experimental Botany*, 35, 239–44.

Kobayashi, M., Tase, K., Yukawa, T. & Egara, K. (1993). Breeding Italian ryegrass for a heavy snowfall area in Japan, improving the two aspects of snow endurance. In *Proceedings XVII International Grassland Congress*, pp. 664–5.

Koblet, R. (1979). Über den bestandsaufbau und die ertragsbildung in dauerwiesen des Alpenraumes. *A. Acker-und Pflanzenbau*, 148, 131–55.

Kolb, D. (1984). *Experiential Learning: Experience as the Source of Learning and Development*. New Jersey: Prentice Hall.

Korte, C.J. (1993). Climate change – adapting the systems and grasslands role. In *Proceedings of XVII International Grassland Congress*, pp. 1247–8.

Kouchi, H. & Nakaji, K. (1985). Utilization and metabolism of photoassimilated ^{13}C in soybean roots and nodules. *Soil Science and Plant Nutrition*, 31, 323–34.

Kowithayakorn, L. & Humphreys, L.R. (1987). Influence of water stress on flowering and seed production of *Macroptilium atropurpureum* cv. Siratro. *Annals of Botany*, 59, 551–7.

Kramer, H.K. (1952). Agronomic problems of strain evaluation with forage crops. In *Proceedings VI International Grassland Congress*, pp. 341–6.

Krieg, N.R. (ed.) (1984). *Bergey's Manual of Systematic Bacteriology*. Baltimore and London: Williams and Wilkins.

Kunelius, H.T., Kim, D.A., Hirota, H. & Zhu, T. (1993). Factors required to sustain pastoral farming systems and forage supply in winter cold zones. In *Proceedings XVII International Grassland Congress*, pp. 651–6.

Laidlaw, A.S. & Reed, K.F.M. (1993). Plant improvement: the evaluation and extension processes. In *Proceedings VI International Grassland Congress*, pp. 385–92.

Lal, R. (1990). *Soil Erosion in the Tropics. Principles and Management*. New York: McGraw-Hill.

Lal, R. & Miller, F.P. (1993). Soil quality and its management in humid subtropical and tropical environments. In *Proceedings XVII International Grassland Congress*, pp. 1541–50.

Lambert, M.G. & Guerin, H. (1990). Competitive and complementary effects with different species of herbivores in their utilisation of pastures. In *Proceedings XVI International Grassland Congress*, Vol. 3, pp. 1785–9.

Langer, R.H.M. (1963). Tillering in herbage grasses. *Herbage Abstracts*, 33, 141–8.

Lawrence, P.A., Cowie, B.A., Yule, D. & Thorburn, P.J. (1993). Water balance and soil fertility characteristics of brigalow (*Acacia harpophylla*) lands before and after forest clearing. In *Proceedings XVII International Grassland Congress*, pp. 2242–4.

Laycock, W.A. (1991). Stable states and thresholds of range condition on North American rangelands: a viewpoint. *Journal of Range Management*, **44**, 427–33.

Lazenby, A. (1983). Nitrogen relationships in grassland ecosystems. In *Proceedings XIV International Grassland Congress*, pp. 56–63.

Leach, G.J. (1970). Growth of the lucerne plant after defoliation. In *Proceedings XI International Grassland Congress*, pp. 562–6.

Leafe, E.L. & Parsons, A.J. (1983). Physiology of growth of a grazed sward. In *Proceedings XIV International Grassland Congress*, pp. 403–6.

Leaver, J.D. (1982). Grass height as a indicator for supplementary feeding of continuously stocked dairy cows. *Grass and Forage Science*, **37**, 285–90.

Lendon, C., Lamacraft, R.R. & Osmond, G. (1976). Standards for testing and assessing range condition in central Australia. *Australian Rangeland Journal*, **1**, 40–8.

Leng, R.A. (1986). Determining the nutritive value of forage. In *Forages in Southeast Asian and South Pacific Agriculture*, ed. G.J. Blair, D.A. Ivory & T.R. Evans, pp. 111–23. Canberra: Australian Centre for International Agricultural Research.

Leng, R.A. (1993). Quantitative ruminant nutrition – a green science. *Australian Journal of Agricultural Research*, **44**, 363–80.

Lenné, J.M. & Trutman, P. (eds.) (1994). *Diseases of Tropical Pasture Plants*. Wallingford, UK: CAB International.

Leukel, W.A. (1937). Growth behaviour and relative composition of pasture grasses as affected by agricultural practices. In *Report IV International Grassland Congress*, pp. 183–6.

Levitt, J.V. (1972). *Responses of Plants to Environmental Stresses*. New York: Academic Press.

Lewis, G.C. & Thomas, B.J. (1991). Incidence and severity of pest and disease damage to white clover foliage at 16 sites in England and Wales. *Annals of Applied Biology*, **118**, 1–8.

Lieth, H.F.H. (ed.) (1978). *Patterns of Primary Productivity in the Biosphere*. Stroudsberg: Hutchinson Ross.

Little, D.A. (1982). Utilization of minerals. In *Nutritional Limits to Animal Production from Pasture*, ed. J.B. Hacker, pp. 259–83. Farnham Royal, UK: Commonwealth Agricultural Bureaux.

Littler, J.W. (1984). Effect of pasture on subsequent wheat crops on a black earth soil of the Darling Downs. 1. The overall experiment. *Queensland Journal of Agricultural and Animal Sciences*, **41**, 1–12.

Lloyd, A.G. (1966). Economic aspects of stocking and feeding policies in the sheep industry in southern Australia. *Proceedings Australian Society of Animal Production*, **6**, 136–47.

Lodge, G.M. & Whalley, R.B.D. (1985). The manipulation of species composition of natural pastures by grazing management on the northern slopes of New South Wales. *Australian Rangeland Journal*, **7**, 6–16.

Loneragan, J.F. & Dowling, E.J. (1958). The interaction of calcium and hydrogen ions in the nodulation of subterranean clover. *Australian Journal of Agricultural Research*, **9**, 464–72.

Loneragan, J.F., Meyer, D., Fawcett, R.G. & Anderson, A.J. (1955). Lime pelleted clover seeds for nodulation on acid soils. *Journal of the Australian Institute of Agricultural Science*, **21**, 264–5.

Long S.P., Garcia Moya E., Imbaba, S.K., Kamnalrut, A., Piedade, M.T.F., Scurlock, J.M.O., Shen, Y.K. & Hall, D.O. (1989). Primary productivity of natural grass ecosystems of the tropics. A reappraisal. *Plant and Soil*, **115**, 155–66.

Lonsdale, W.M. (1994). Inviting trouble: introduced pasture species in northern Australia. *Australian Journal of Ecology*, **19**, 345–54.

Lorenzetti, F. (1993). Achieving potential herbage seed yields in species of temperate regions. In *Proceedings XVII International Grassland Congress*, pp. 1621–8.

Love, R.M. (1952). The value of induced polyploidy in breeding. In *Proceedings VI International Grassland Congress*, pp. 292–8.

Lowe, J. (1966). Output of pastures under a clover nitrogen regime in Northern Ireland. In *Proceedings X International Grassland Congress*, pp. 187–91.

Lowe, J. (1970). Comparative efficiency of pastures and crops for animal production. In *Proceedings XI International Grassland Congress*, pp. A88–94.

Ludlow, M.M. (1980). Stress physiology of tropical pasture plants. *Tropical Grasslands*, **14**, 136–45.

Ludlow, M.M. (1989). Strategies of response to water stress. In *Structural and Functional Responses to Environmental Stress: Water Shortage*, ed. K.H. Kreeb, M. Richter & T.M. Hanckley, pp. 269–81. The Hague: SBP Academic Publishing.

Ludlow, M.M. & Björkman, O. (1984). Paraheliotropic leaf movement in Siratro as a protective mechanism against drought-induced damage to primary photosynthetic reactions: damage by excessive light and heat. *Planta*, **161**, 505–18.

Ludlow, M.M. & Charles-Edwards, D.A. (1980). Analysis of the regrowth of a tropical grass/legume sward subjected to different frequencies and intensities of defoliation. *Australian Journal of Agricultural Research*, **31**, 673–92.

Ludlow, M.M. & Wilson, G.L. (1971a). Photosynthesis of tropical pasture plants. 1. Illuminance, carbon dioxide concentration, leaf temperature, and leaf–air vapour pressure difference. *Australian Journal of Biological Sciences*, **24**, 449–70.

Ludlow, M.M. & Wilson, G.L. (1971b). Photosynthesis of tropical pasture plants. 2. Temperature and illuminance history. *Australian Journal of Biological Sciences*, **24**, 1065–75.

Ludwig, J.A., Clewett, J.F. & Foran, B.D. (1993). Meeting the needs of decision support system users. In *Decision Support Systems for the Management of Grazing Lands*, ed. J.W. Stuth & B.G. Lyons, pp. 209–20. Carnforth, UK: UNESCO Man and the Biosphere Book 11, Parthenon Publishing.

McCall, D.G., Marshall, P.R. & Johns, K.L. (1991). An introduction to STOCKPOL: a decision support system for livestock farms. In *Proceedings of the International Conference on Decision Support Systems for Resource Management*, pp. 27–30.

McCall, D.G. & Sheath, G.W. (1993). Development of intensive grassland systems: from science to practice. In *Proceedings XVII International Grassland Congress*, pp. 1257–65.

McClintock, D. & Ison, R. (1994). Revealing and concealing metaphors for a systemic agriculture. In *Systems-Oriented Research in Agriculture and Rural Development*, ed. M. Sebillotte, pp. 212–16. Montpellier: CIRAD-SAR.

McCosker, T. (1994). The dichotomy between research results and practical experience with time control grazing. In *Australian Rural Science Annual 1994*, pp. 26–31. Armidale: University of New England.

McCracken, J., Pretty, J. & Conway, G. (1988). *An Introduction to Rapid Rural Appraisal for Agricultural Development*. London: IIED.

McDowell, L.R., Conrad, J.H., Ellis, G.L. & Loosli, J.K. (1983). *Minerals for Grazing Ruminants in Tropical Regions.* Gainesville, Florida: University of Florida.
McDowell, L.R., Conrad, J.H., Thomas, J.E., Harris, L.E. & Fick, K.R. (1977). Nutritional composition of Latin American forages. *Tropical Animal Production,* **2**, 273–9.
McGinnies, W.J. & Nicholas, P.J. (1983). Effects of topsoil depths and species selection on reclamation of coal-strip-mine spoils. In *Proceedings XIV International Grassland Congress,* pp. 353–6.
McIntyre, C.L., Manners, J.M., Wilson, J.R., Way, H. & Sharp, D. (1993). Genetic engineering of pasture legumes and grasses for reduced lignin content and increased digestibility. In *Proceedings XVII International Grassland Congress,* pp. 1100–2.
McIvor, J.G. (1984). Leaf growth and senescence in *Urochloa mosambicensis* and *U. oligotricha* in a seasonally dry tropical environment. *Australian Journal of Agricultural Research,* **35**, 177–87.
McIvor, J.G. (1993). Distribution and abundance of plant species in pastures and rangelands. In *Proceedings XVII International Grassland Congress,* pp. 285–90.
McIvor, J.G., Ash, A.J. & Cook, G.D. (1995). Land condition in the tropical tallgrass pasture lands. I. Effects on herbage production. *The Rangeland Journal,* **17**, 69–85.
McIvor, J.G. & Scanlan, J.C. (1994). State and transition models for rangelands. 8. A state and transition model for the northern speargrass zone. *Tropical Grasslands,* **28**, 256–9.
McKenzie, F.R. & Tainton, N.M. (1993). Volatilised nitrogen from a tropical *Pennisetum clandestinum* in South Africa. In *Proceedings XVII International Grassland Congress,* pp. 1563–5.
McKeon, G.M., Day, K.A., Howden, S.M., Mott, J.J., Orr, D.M., Scattini, W.J. & Weston, E.J. (1990). Management for pastoral production in northern Australian savannas. *Journal of Biogeography,* **17**, 355–72.
McKeon, G.M., Howden, S.M., Abel, N.O.J. & King, J.M. (1993). Climate change: adapting tropical and subtropical grasslands. In *Proceedings XVII International Grassland Congress,* pp. 1181–90.
McMeekan, C.P. (1952). Interdependence of grassland and livestock in agricultural production. In *Proceedings VI International Grassland Congress,* pp. 149–61.
McMeekan, C.P. (1956). Grazing management and animal production. In *Proceedings VII International Grassland Congress,* pp. 146–56.
McMeekan, C.P. (1961). Grazing Management. In *Proceedings VIII International Grassland Congress,* pp. 21–6.
McNaughton, S.J. (1985). Ecology of a grazing system: the Serengeti. *Ecological Monographs,* **55**, 259–94.
Mannetje, L.'t. & Ebersohn, J.P. (1980). Relations between sward characteristics and animal production. *Tropical Grasslands,* **14**, 273–87.
Marshall, D.R. & Brown, A.D.H. (1983). Theory of forage plant collection. In *Genetic Resources of Forage Plants,* ed. J.G. McIvor & R.A. Bray, pp. 135–48. Melbourne: CSIRO.
Mathison, M.J. (1983). Mediterranean and temperate forage legumes. In *Genetic Resources of Forage Plants,* ed. J.G. McIvor & R.A. Bray, pp. 63–84. Melbourne: CSIRO.
Maxwell, T.J., Sibbald, A.R., Dalziel, A.J.I., Agnew, R.D.M. & Elston, D.A.

(1994). The implications of controlling grazed sward height for the operation and productivity of upland sheep systems in the UK. I. Effects of two annual stocking rates in combination with two sward height profiles. *Grass and Forage Science*, **49**, 73–88.

May, L.H. & Davidson, J.L. (1958). The role of carbohydrate reserves in regeneration of plants. 1. Carbohydrate changes in subterranean clover following defoliation. *Australian Journal of Agricultural Research*, **9**, 767–77.

Mayer, A. (1987). *Data Base in Forages of the Mediterranean Basin and Adjacent Semi-Arid Areas*. Rome: IBPGR Report 88/154.

Mayne, C.S., Newberry, R.D., Woodcock, S.C.F. & Wilkins, R.J. (1987). Effect of grazing severity on grass utilization and milk production of rotationally grazed dairy cows. *Grass and Forage Science*, **42**, 59–72.

Mays, D.A. (1970). Sulphur-coated urea: a slow-release nitrogen source for grass. In *Proceedings XI International Grassland Congress*, pp. 428–30.

Mears, P.T. & Humphreys, L.R. (1974a). Nitrogen response and stocking rate of *Pennisetum clandestinum* pastures. I. Pasture nitrogen requirement and concentration, distribution of dry matter and botanical composition. *Journal of Agricultural Science*, **83**, 451–68.

Mears, P.T. & Humphreys, L.R. (1974b). Nitrogen response and stocking rate of *Pennisetum clandestinum* pastures. II. Cattle growth. *Journal of Agricultural Science*, **83**, 469–78.

Mehra, K.L. & Magoon, M.L. (1974). Gene centres of tropical and sub-tropical pasture legumes and their significance in plant introductions. In *Proceedings XII International Grassland Congress*, p. 251.

Michaud, R., Viands, D.R. & Christie, B.R. (1993). Comparison of polycross and topcross progeny testing in alfalfa. In *Proceedings XVII International Grassland Congress*, pp. 412–13.

Minami, K., Goudriaan, J., Lentinga, E.A. & Kimura, T. (1993). Significance of grasslands in emission and absorption of greenhouse gases. In *Proceedings XVII International Grassland Congress*, pp. 1231–8.

Minchin, F. (1994). The time course of oxygen diffusion barrier operation in legume nodules. In *1993 IGER Report*, pp. 46–7. Aberystwyth: AFRC Institute of Grassland and Environmental Research.

Minson, D.J. (1971). The digestibility and voluntary intake of six *Panicum* varieties. *Australian Journal of Experimental Agriculture and Animal Husbandry*, **11**, 18–25.

Minson, D.J. (1990). *Forage in Ruminant Nutrition*. San Diego: Academic Press.

Minson, D.J., Raymond, W.F. & Harris, C.E. (1960). The digestibility of grass species and varieties. In *Proceedings VIII International Grassland Congress*, pp. 470–4.

Minson, D.J. & Wilson, J.R. (1994). Prediction of intake as an element of forage quality. In *Forage Quality, Evaluation and Utilization*, ed. G.C. Fahey Jr, M. Collins, D.R. Mertens & L.E. Moser, pp. 533–63. Madison, WI: American Society of Agronomy.

Mitchell, N.R. (1993). A traditional exporter's perspective of trade reforms. In *Proceedings XVII International Grassland Congress*, pp. 981–3.

Moody, M.E. & Mack, R.N. (1988). Controlling the spread of plant invasions: the importance of nascent foci. *Journal of Applied Ecology*, **25**, 1009–21.

Mooney, H.A. (1993). Human impact on terrestrial ecosystems – what we know and what we are doing about it. In *Proceedings XVII International Grassland Congress*, pp. 11–14.

Moore, R.M. (1967). Interaction between the grazing animal and its environment.

In *Report of Proceedings and Invited Papers, 9th International Congress of Animal Production, 1966*, pp. 188–95. Edinburgh: Oliver & Boyd.

Moore, R.M., Barrie, N. & Kipps, E.H. (1946). *Grazing Management: Continuous and Rotational Grazing by Merino Sheep. A Study of the Production of a Sown Pasture in the ACT under Three Systems of Grazing Management*, CSIR Bulletin 201. Melbourne: CSIR.

Morris, S.T., Hirschberg, S.W., Michel, A., Parker, W.J. & McCutcheon, S.N. (1993). Herbage intake and liveweight gain of bulls and steers continuously stocked at fixed sward heights during autumn and spring. *Grass and Forage Science*, **48**, 109–17.

Mott, G.O. (1961). Grazing pressure and the measurement of pasture production. In *Proceedings VIII International Grassland Congress*, pp. 606–11.

Mott, G.O. & Lucas, H.L. (1952). The design, conduct and interpretation of grazing trials on cultivated and improved pastures. In *Proceedings VI International Grassland Congress*, pp. 1380–5.

Mott, J.J. (1985). Mosaic grazing – animal selectivity in tropical savannas of northern Australia. In *Proceedings XV International Grassland Congress*, pp. 1129–30.

Mott, J.J., Ludlow, M.M., Richards, J.H., Parsons, A.D. (1992). Effects of moisture supply in the dry season and subsequent defoliation on persistence of the savanna grasses *Themeda triandra, Heteropogon contortus* and *Panicum maximum*. *Australian Journal of Agricultural Research*, **43**, 241–60.

Mott, J.J., McKeon, G.K. & Day, K.A. (1993). Prediction of plant mortality under grazing – a conceptual approach. In *Proceedings XVII International Grassland Congress*, pp. 167–8.

Mulder, E.G. (1952). Fertilizer vs. legume nitrogen for grasslands. In *Proceedings VI International Grassland Congress*, pp. 740–8.

Muldoon, D.K. & Pearson, C.J. (1979a). Primary growth and re-growth of the tropical tallgrass hybrid *Pennisetum* at different temperatures. *Annals of Botany*, **43**, 709–17.

Muldoon, D.K. & Pearson, C.J. (1979b). Morphology and physiology of re-growth of the tropical tallgrass hybrid *Pennisetum*. *Annals of Botany*, **43**, 719–28.

Murtagh, G.J. (1975). Environmental effects on the short-term response of tropical grasses to nitrogen fertilizer. *Australian Journal of Experimental Agriculture and Animal Husbandry*, **15**, 679–88.

Murtagh, G.J. (1977). The climate induced probability distribution of short-term responses of a tropical grass to nitrogen fertilizer. *Australian Journal of Experimental Agriculture and Animal Husbandry*, **17**, 614–20.

Murtagh, G.J. & Gross, H.D. (1966). Interception of solar radiation and growth rate of a grass sward. In *Proceedings X International Grassland Congress*, pp. 104–8.

Murtagh, G.J., Kaiser, A.G., Huett, D.O. & Hughes, R.M. (1980). Summer-growing components of a pasture system in a subtropical environment. I. Pasture growth, carrying capacity and milk production. *Journal of Agricultural Science*, **94**, 645–63.

Mytton, L.R. & Skøt, L. (1993). Breeding for improved symbiotic nitrogen fixation. In *Plant Breeding: Principles and Prospects*, ed. M.D. Hayward, N.O. Bosemark & I. Ramagoza, pp. 451–73. London: Chapman & Hall.

National Greenhouse Gas Inventory Committee (1994). *Summary – Australian Methodology for the Estimation of Greenhouse Gas Emissions and Sinks and National Greenhouse Gas Inventory 1988 and 1990*. Canberra: Department of the Environment, Sports and Territories.

Neller, J.R. (1952). Sulfur in relation to fertilizers and soil amendments for grasslands. In *Proceedings VI International Grassland Congress*, pp. 728–34.

Newman, J.A., Parsons, A.J. & Harvey, A. (1992). Not all sheep prefer clover: diet selection revisited. *Journal of Agricultural Science*, 119, 275–83.

Newton, P.C.D., Clark, H., Bell, C.C., Glascow, E.M. & Campbell, B.D. (1994). Effect of elevated CO_2 and simulated seasonal changes in temperature on the species composition and growth rate of pasture turves. *Annals of Botany*, 73, 53–9.

Nicholson, C.F., Lee, D.R., Boisvert, R.N., Blake, R.W. & Urbina, C.I. (1994). An optimization model of the dual-purpose cattle production system in the humid lowlands of Venezuela. *Agricultural Systems*, 46, 311–34.

Nie, D., He, H., Kirkham, M.B. & Kanemasu, E.T. (1993). Photosynthesis and water relations of a C_4 and a C_3 grass under doubled carbon dioxide. In *Proceedings XVII International Grassland Congress*, pp. 1139–41.

Nojima, H., Oizumi, H. & Takasaki, Y. (1985). Effect of cytokinin on lateral bud development in regrowth of *Sorghum bicolor* M. In *Proceedings XV International Grassland Congress*, pp. 372–3.

Norby, R.J. (1987). Nodulation and nitrogenase activity in nitrogen fixing woody plants stimulated by CO_2 enrichment of the atmosphere. *Physiologia Plantarum*, 71, 77–82.

Nores, G.A. & Vera, R.R. (1993). Science and information for our grasslands. In *Proceedings XVII International Grassland Congress, Palmerston North, NZ*, pp. 33–7.

Norris, D.O. (1956). Legumes and the *Rhizobium* symbiosis. *Empire Journal of Experimental Agriculture*, 24, 247–70.

Norris, D.O. (1967). The intelligent use of inoculants and lime pelleting for tropical legumes. *Tropical Grasslands*, 1, 107–21.

Norris, K.H., Barnes, R.F., Moore, J.E. & Shenk, J.S. (1976). Predicting forage quality by infrared reflectance spectroscopy. *Journal of Animal Science*, 43, 889–97.

Norton, B.W. (1994). The nutritive value of tree legumes. In *Forage Tree Legumes in Tropical Agriculture* ed. R.C. Gutteridge & H.M. Shelton, pp. 177–91. Wallingford, UK: CAB International.

Norton, B.W. & Hales, J.W. (1976). A response of sheep to cobalt supplementation in south-eastern Queensland. *Proceedings of the Australian Society of Animal Production*, 11, 393–6.

Noy-Meir, I. (1980). Structure and function of desert ecosystems. *Israel Journal of Botany*, 28, 1–19.

Nutman, P.S. (1957). Studies on the physiology of nodule formation. V. Further experiments on the stimulating and inhibitory effects of root secretions. *Annals of Botany*, 21, 321–7.

Oberle, S. (1994). Farming systems options for US agriculture: an agroecological perspective. *Journal of Production Agriculture*, 7, 119–23.

O'Connor, T.G. (1985). A synthesis of field experiments concerning the grass layer in the savanna regions of southern Africa. South African National Scientific Programmes Report No. 114, South Africa: Foundation for Research Development.

Oizumi, H., Takasaki, Y., Nojima, H. & Isono, Y. (1985). Physiology of regrowth in *Sorghum bicolor* Moench: studies on the sequential action of hormones and reserves. In *Proceedings XV International Grassland Congress*, 419–21.

Olmsted, C.E. (1952). Photoperiodism in native range grasses. In *Proceedings VI International Grassland Congress*, pp. 676–82.

O'Reagain, P.J. & Turner, J.R. (1992). An evaluation of the empirical basis for grazing management recommendations for rangeland in southern Africa. *Journal of the Grassland Society of Southern Africa*, 9, 38–49.

Orpin, C.G. & Xue, G. (1993). Genetics of fibre degradation in the rumen, particularly in relation to anaerobic fungi, and its modification by recombinant DNA technology. In *Proceedings XVII International Grassland Congress*, pp. 1209–14.

Orr, D.M. (1980). Effects of sheep grazing *Astrebla* grasslands in central western Queensland. I. Effects of grazing pressure and livestock distribution. *Australian Journal of Agricultural Research*, 31, 797–806.

Orr, D.M., McKeon, G.M. & Day, K.A. (1991). Burning and exclosure can rehabilitate degraded black spear grass (*Heteropogon contortus*) pastures. *Tropical Grasslands*, 25, 333–6.

Orr, R.J., Parsons, A.J., Penning, P.D. & Treacher, T.T. (1990). Sward composition, animal performance and the potential production of grass/white clover swards continuously stocked with sheep. *Grass and Forage Science*, 45, 325–36.

Orr, R.J., Parsons, A.J., Treacher, T.T. & Penning, P.D. (1988). Seasonal patterns of grass production under cutting or continuous stocking management. *Grass and Forage Science*, 43, 199–207.

Orr, R.J., Penning, P.D., Parsons, A.J. & Champion, R.A. (1995). Herbage intake and N excretion by sheep grazing monocultures or a mixture of grass and white clover. *Grass and Forage Science*, 50, 31–40.

Osty, P.L. & Landais, E. (1993). Functioning of pastoral farming systems. In *Systems Studies in Agriculture and Rural Development*, ed. J. Brossier, L. de Bonneval & E. Landais, pp. 201–13. Paris: INRA.

Osvald, H. (1937). Achievements and aims in modern Swedish grassland management. In *Report IV International Grassland Congress*, pp. 46–53.

Othman, W.M.W., Asher, C.J. & Wilson, G.L. (1988). ^{14}C-labelled assimilate supply to root nodules and nitrogen fixation of phasey bean plants following defoliation and flower removal. In *Biotechnology of Nitrogen Fixation in the Tropics*, ed. Z.H. Shamsuddin, W.M.W. Othman, M. Marziah & J. Sundram, pp. 217–24. Kuala Lumpur: Universiti Pertanian Malaysia.

Ourry, A., Boucaud, J. & Salette, J. (1988). Nitrogen mobilization from stubble and roots during re-growth of defoliated perennial rye grass. *Journal of Experimental Botany*, 39, 803–9.

Owensby, C.E. (1993). Climate change and grasslands: ecosystem-level responses to elevated carbon dioxide. In *Proceedings XVII International Grassland Congress*, pp. 1119–24.

Packham, R.G., Ison, R.L. & Roberts, R.J. (1988). Soft-systems methodology for action research: the role of a college farm in an agricultural education institution. *Agricultural Administration and Extension*, 30, 109–26.

Parker, K.W. (1952). The role of plant ecology in range research and range management. In *Proceedings VI International Grassland Congress*, pp. 618–24.

Parker, W.J. & McCutcheon, S.N. (1992). Effect of sward height on herbage intake and production of ewes of different rearing rank during lactation. *Journal of Agricultural Science*, 118, 383–95.

Parry, M. & Rosenzweig, C. (1993). Implications for grasslands of the response to climate change by the world food system. In *Proceedings XVII International Grassland Congress*, pp. 1239–46.

Parsons, A.J. (1993). Chairperson's summary paper. Session 8: plant growth. In *Proceedings XVII International Grassland Congress*, pp. 176–8.

Parsons, A.J., Harvey, A. & Woledge, J. (1991). Plant–animal interactions on a continuously grazed mixture. 1. Differences in the physiology of leaf expansion and the fate of leaves of grass and clover. *Journal of Applied Ecology*, **28**, 619–34.

Parsons, A.J., Harvey, A. & Johnson, I.R. (1991). Plant/animal interactions in continuously grazed mixtures. 2. The role of differences in the physiology of plant growth and of selective grazing on the performance and stability of species in a mixture. *Journal of Applied Ecology*, **28**, 635–58.

Parsons, A.J., Johnson, I.R. & Harvey, A. (1988). Use of a model to optimize the interaction between frequency and severity of intermittent defoliation and to provide a fundamental comparison of the continuous and intermittent defoliation of grass. *Grass and Forage Science*, **43**, 49–59.

Parsons, A.J., Johnson, I.R. & Williams, J.H.H. (1988). Leaf age structure and canopy photosynthesis in rotationally and continuously grazed swards. *Grass and Forage Science*, **43**, 1–14.

Parsons, A.J., Leafe, E.L., Collett, B. & Stiles, W. (1983a). The physiology of grass production under grazing. 1. Characteristics of leaf and canopy photosynthesis of continuously-grazed swards. *Journal of Applied Ecology*, **20**, 117–26.

Parsons, A.J., Leafe, E.L., Collett, B. & Stiles, W. (1983b). The physiology of grass production under grazing. 2. Photosynthesis, crop growth and animal intake of continuously grazed swards. *Journal of Applied Ecology*, **20**, 127–40.

Parsons, A.J. & Penning, P.D. (1988). The effect of the duration of regrowth on photosynthesis, leaf death and the average rate of growth in a rotationally grazed sward. *Grass and Forage Science*, **43**, 15–27.

Parsons, A.J. & Robson, M.J. (1982). Seasonal changes in the physiology of S24 perennial ryegrass (*Lolium perenne* L.). 4. Comparison of the carbon balance of the reproductive crop in spring and the vegetative crop in autumn. *Annals of Botany*, **50**, 167–77.

Partridge, I.J. & Ranacou, E. (1974). The effects of supplemental *Leucaena leucocephala* browse on steers grazing *Dichanthium caricosum* in Fiji. *Tropical Grasslands*, **8**, 107–12.

Pasumarty, S.V., Matsumura, T., Higuchi, S. & Yamada, T. (1993). Ovule fertility – a tool for selecting high-fertility populations of white clover. In *Proceedings XVII International Grassland Congress*, pp. 1648–9.

Patil, B.D. & Jones, D.I.H. (1970). The mineral status of some temperate herbage varieties in relation to animal performance. In *Proceedings XI International Grassland Congress*, pp. 726–30.

Paul, E.A. (1988). Towards the year 2000: directions for future nitrogen research. In *Advances in Nitrogen Cycling in Agricultural Ecosystems*, ed. J.R. Wilson, pp. 417–25. Wallingford, UK: CAB International.

Peacock, W.J. (1993). Genetic engineering for pastures. In *Proceedings XVII International Grassland Congress*, pp. 29–32.

Pearcy, R.W. (1987). Photosynthetic responses of tropical forest trees. In *Proceedings International Conference on Tropical Plant Ecophysiology* December 4–6, 1985, Bogor, Indonesia, pp. 49–66. Biotrop Special Publication 31.

Pearman, G.I. (ed.) (1988). *Greenhouse: Planning for Climatic Change*. Melbourne: CSIRO. Leiden: E.J. Brill.

Pearson, C.J. & Ison, R.L. (1987). *Agronomy of Grassland Systems*. Cambridge: Cambridge University Press.

Pengelly, B.C. & Williams, R.J. (1993). Improved methods of selecting tropical germplasm for plant breeding and species evaluation studies. In *Proceedings XVII International Grassland Congress*, pp. 225–6.

Penning, P.D., Parsons, A.J., Orr, R.J. & Treacher, T.T. (1991). Intake and behaviour responses by sheep to changes in sward characteristics under continuous stocking. *Grass and Forage Science*, **46**, 15–28.

Peters, N.K., Frost, J. & Long, S.R. (1986). A plant flavone, luteolin, induces expression of *Rhizobium meliloti* nodulation genes. *Science*, **23**, 977–80.

Petersen, R.G., Lucas, H.L. & Mott, G.O. (1965). Relationship between rate of stocking and per animal and per acre performance on pasture. *Agronomy Journal*, **57**, 27–30.

Petty, S., Groot, T. & Triglone, T. (1994). *Leucaena Production in the Ord River Irrigation Area*. Farmnote 106/94. Perth: Western Australian Department of Agriculture.

Phillips, I.D.J. (1975). Apical dominance. *Annual Review of Plant Physiology*, **26**, 341–67.

Phular, J.J., Knight, R.W. & Heitschmidt, R.K. (1987). Infiltration rates and sediment production as influenced by grazing systems in the Texas Rolling Plains. *Journal of Range Management*, **40**, 240–3.

Picard, J. & Desroches, R. (1976). Essai d'estimation des surfaces cultivés en France en graminées et legumineuses prairiales. *Fourrages*, **66**, 3–30.

Picket, S.T.A. & White, P.S. (eds.) (1985). *The Ecology of Natural Disturbance and Patch Dynamics*. New York: Academic Press.

Pickup, G. (1989). New land degradation survey techniques for arid Australia – problems and prospects. *Australian Rangeland Journal*, **11**, 74–82.

Pieper, R.D. & Heitschmidt, R.K. (1988). Is short-duration grazing the answer? *Journal of Soil and Water Conservation*, **43**, 133–7.

Pilbeam, C.J. & Robson, M.J. (1992). Responses of populations of *Lolium perenne* cv. S23 with contrasting rates of dark respiration to nitrogen supply and defoliation regime. 1. Grown as monocultures. *Annals of Botany*, **69**, 69–77.

Pinheiro, S.L.G., Pearson, C.J. & Ison, R.L. (1994). A farming systems research/extension model underway in Santa Catarina, Brazil: a critical analysis. In *Systems-Oriented Research in Agriculture and Rural Development*, ed. M. Sebillotte, p. 2804. Montpellier: CIRAD-SAR.

Pittock, A.B. (1993). A climate change perspective on grasslands. In *Proceedings XVII International Grassland Congress*, pp. 1053–60.

Playne, M.J., McLeod, M.N. & Dekker, R.F.H. (1972). Digestion of dry matter, nitrogen, phosphorus, sulphur, calcium and detergent fibre fractions of the seed and pod of *Stylosanthes humilis* contained in terylene bags in the bovine rumen. *Journal of Science, Food and Agriculture*, **23**, 925–32.

Pleasants, A.B., McCall, D.G. & Sheath, G.W. (1995). Management of pastoral grazing systems under biological and environmental variation. *Agricultural Systems*, **48**, 179–92.

Poppi, D.P. & McLennan, S.R. (1995). Protein and energy utilization by ruminants at pasture. *Journal of Animal Science* **73**, 278–90.

Probert, M.E., Okalebo, J.R. & Jones, R.K. (1995). The use of manure on smallholders' farms in semi-arid eastern Kenya. *Experimental Agriculture*, **31**, 371–81.

Pulsford, J.S. (1980). Trends in fertilizer costs and usage on pastures in tropical and subtropical Australia. *Tropical Grasslands*, **14**, 188–93.

Purvis, O.N. (1952). The physiology of flower differentiation in grasses. In *Proceedings VI International Grassland Congress*, pp. 661–6.

Quirk, M.F., Bushell, J.J., Jones, R.J., Megarrity, R.C. & Butler, K.L. (1988).

Liveweight gains on leucaena and native grass pastures after dosing cattle with rumen bacteria capable of degrading DHP, a ruminal metabolite from leucaena. *Journal of Agricultural Science*, 111, 165–70.

Quirk, M.F., Paton, C.J. & Bushell, J.J. (1990). Increasing the amount of leucaena on offer gives faster growth rates of grazing cattle in South East Queensland. *Australian Journal of Experimental Agriculture*, 30, 51–4.

Rawes, M. (1981). Further results of excluding sheep from high-level grasslands in the north Pennines. *Journal of Ecology*, 69, 651–69.

Reddell, P., Rosbrook, P.A., Bowen, G.D. & Gwaze, D. (1988). Growth responses in *Casuarina cunninghamiana* plantings to inoculation with *Frankia. Plant and Soil*, 108, 79–86.

Reich, T.J., Iyer, V.H. & Miki, B.L. (1986). Efficient transformation of alfalfa protoplasts by the intranuclear microinjection of Ti plasmids. *Biotechnology*, 4, 1001–4.

Reid, R. (1993). Establishing and sharing collections of a valuable global resource. In *Proceedings XVII International Grassland Congress*, pp. 189–94.

Reid, R., Konopka, J., Perret, P.M. & Guarino, L. (1989). Forage germplasm resources for the Mediterranean and adjacent arid/semi-arid areas. In *Proceedings XVI International Grassland Congress*, pp. 289–90.

Reid, R. & Strickland, R.W. (1983). Forage plant collection in practice. In *Genetic Resources of Forage Plants*, ed. J.G. McIvor & R.A. Bray, pp. 149–56. Melbourne: CSIRO.

Reid, R.L. (1994). Milestones in forage research (1969–1994). In *Forage Quality, Evaluation and Utilization*, ed. G.C. Fahey Jr, M. Collins, D.R. Mertens & L.E. Moser, pp. 1–58. Madison, WI: American Society of Agronomy.

Reid, R.L., Jung, G.A., Roemig, I.J. & Kocher, R.E. (1978). Mineral utilization by lambs and guinea pigs fed Mg fertilized grass. *Agronomy Journal*, 70, 9–14.

Rerkasem, K. & Rerkasem, B. (1988). Yield and nitrogen nutrition of intercropped maize and ricebean (*Vigna umbellata*). *Plant and Soil*, 108, 151–62.

Rhodes, I. (1973). Relationship between canopy structure and productivity in herbage grasses and its implications for plant breeding. *Herbage Abstracts*, 43, 129–33.

Rhodes, I. (1975). The relationships between productivity and some components of canopy structure in ryegrass (*Lolium perenne*). IV. Canopy characters and their relationship with sward yields in some intra population selections. *Journal of Agricultural Science*, 84, 345–51.

Rhodes, I. & Webb, K.J. (1993). Improvement of white clover. *Outlook on Agriculture*, 22, 189–94.

Rhodes, I., Webb, K.J., Evans, D.R. & Collins, R.P. (1993). Problems, potentialities and progress in white clover breeding. In *Proceedings XVII International Grassland Congress*, pp. 443–5.

Riceman, D.S. (1952). Minor element deficiencies and their correction. In *Proceedings VI International Grassland Congress*, pp. 710–17.

Richards, J.H. (1984). Root growth response to defoliation in two *Agropyron* bunchgrasses: field observations with an improved root periscope. *Oecologia*, 64, 21–5.

Richards, J.H. (1993). Physiology of plants recovering from defoliation. In *Proceedings XVII International Grassland Congress*, pp. 85–93.

Richards, J.H. & Caldwell, M.M. (1985). Soluble carbohydrates, concurrent photosynthesis and efficiency in regrowth following defoliation: a field study with *Agropyron* species. *Journal of Applied Ecology*, 22, 907–20.

246   References

Riewe, M.E. (1961). Use of the relationship of stocking rate to gain of cattle in an experimental design for grazing trials. *Agronomy Journal*, **53**, 309–13.

Robards, G.E., Michalk, D.L. & Pither, R.J. (1978). Evaluation of natural annual pasture at Trangie in central western New South Wales. 3. Effect of stocking rate on annual-dominated and perennial-dominated pastures. *Australian Journal of Experimental Agriculture and Animal Husbandry*, **18**, 361–9.

Robbins, G.B., Bushell, J.J. & Butler, K.L. (1987). Decline in plant and animal production from ageing pastures of green panic (*Panicum maximum* var. *trichoglume*). *Journal of Agricultural Science*, **108**, 407–17.

Robbins, G.B., Bushell, J.J. & McKeon, G.M. (1989). Nitrogen immobilization in decomposing litter contributes to productivity decline in ageing pastures of green panic (*Panicum maximum* var. *trichoglume*). *Journal of Agricultural Science*, **113**, 401–6.

Roberts, B.R. (1970). Assessment of veld condition and trend. *Proceedings of Grassland Society of Southern Africa*, **5**, 137–9.

Roberts, C.R. (1980). Effect of stocking rate on tropical pastures. *Tropical Grasslands*, **14**, 225–31.

Roberts, E.H. (1972). *Viability of Seeds*. Syracuse: Syracuse University.

Robin, Ch., Varlet-Grancher, C., Gastol, F. & Guckert, A. (1993). Phytochrome control of assimilate partitioning in white clover. In *Proceedings XVII International Grassland Congress*, pp. 159–61.

Roe, R. (1987). Recruitment of *Astrebla* spp. in the Warrego region of south-western Queensland. *Tropical Grasslands*, **21**, 91–2.

Roeckner, E. (1992). Past, present and future levels of greenhouse gases in the atmosphere and model projections of related climatic changes. *Journal of Experimental Botany*, **43**, 1097–109.

Rosenthal, G.A. & Janzen, D.H. (1979). *Herbivores. Their Interaction with Secondary Plant Metabolites*. New York: Academic Press.

Rotili, P. (1993). Selfing and competition applied to lucerne breeding and to variety constitution. In *Proceedings XVII International Grassland Congress*, pp. 397–8.

Rowland, J.W. & Bumpus, E.D. (1952). The place of legumes in the sward. In *Proceedings VI International Grassland Congress*, pp. 468–73.

Russell, D.B. & Ison, R.L. (1992). The research–development relationship in rangelands: an opportunity for contextual science. In *Proceedings IV International Rangeland Congress, Montpellier, France*, pp. 1047–54.

Russell, J.S. (1991). Likely climatic changes and their impact on the northern pastoral industry. *Tropical Grasslands*, **25**, 211–18.

Russell, J.S. & Williams, C.H. (1982). Biogeochemical interactions of carbon, nitrogen, sulfur and phosphorus in Australian agroecosystems. In *Cycling of Carbon, Nitrogen, Sulfur and Phosphorus in Terrestrial and Aquatic Ecosystems*, ed. I.E. Galbally & J.R. Freney. pp. 61–75. Canberra: Australian Academy of Science.

Ruz-Jerez, B.E., Ball, P.R. & White, R.E. (1993). Improving the efficiency of nitrogen utilisation in a grassland soil with different mixtures of pasture species. In *Proceedings XVII International Grassland Congress*, pp. 1439–41.

Ryden, J.C., Ball, P.R. & Garwood, E.A. (1984). Nitrate leaching from grassland. *Nature*, **311**, 50–3.

Ryle, G.A. (1966). Physiological aspects of seed yield in grasses. In *The Growth of Cereals and Grasses*, ed. F.L. Milthorpe, pp. 106–18. London: Butterworths.

Sackville Hamilton, N.R., Chorlton, K.H., Thomas, I.D. & Hayward, M.D. (1993). Germplasm collecting by IGER: the future. In *Proceedings XVII International Grassland Congress*, pp. 219–20.

Saeki, T. (1960). Interrelationships between leaf amount, light distribution and total photosynthesis in a plant community. *Botanical Magazine (Tokyo)*, **73**, 155–63.

Sampson, A.W. (1919). *Plant Succession in Relation to Range Management*. Washington, DC: USDA Bulletin 791.

Sampson, A.W. (1939). Plant indicators – concept and status. *Botanical Review*, **5**, 155–206.

Sampson, A.W. (1952). *Range Management – Principles and Practices*. New York: John Wiley.

Sanginga, N., Zapata, F., Danso, S.K.A. & Bowen, G.D. (1990). Effect of successive cuttings on uptake and partitioning of ^{15}N among plant parts of *Leucaena leucocephala*. *Biology and Fertility of Soils*, **9**, 37–42.

Savory, A. (1978). A holistic approach to range management using short duration grazing. In *Proceedings First International Rangeland Congress*, ed. D.N. Hyder, pp. 555–7.

Savory, A. (1983). The Savory grazing method or holistic resource management. *Rangelands*, **5**, 155–9.

Scanlan, J.C., McKeon, G.M., Day, K.A., Mott, J.J. & Hinton, A.W. (1994). Estimated safe carrying capacities of extensive cattle-grazing properties within tropical, semi-arid woodlands of north-eastern Australia. *The Rangeland Journal*, **16**, 64–76.

Schofield, J.L. (1941). Introduced legumes in north Queensland. *Queensland Agricultural Journal*, **56**, 378–88.

Scholefield, D., Tyson, K.C., Garwood, E.A., Armstrong, A.C., Hawkins, J. & Stone, A.C. (1993). Nitrate leaching from grazed grassland lysimeters: effects of fertilizer input, field drainage, age of sward and patterns of weather. *Journal of Soil Science*, **44**, 601–13.

Schultze-Kraft, R., Williams, W.M. & Keoghan, J.M. (1993). Searching for new germplasm for the year 2000 and beyond. In *Proceedings XVII International Grassland Congress*, pp. 181–7.

Schulz, E.F., Langford, W.R., Evans, E.M., Patterson, R.M. & Anthony, W.B. (1959). Relationship of beef gains to forage yields. *Agronomy Journal*, **51**, 207–11.

Scoones, I. & Thompson, J. (1994). Knowledge, power and agriculture – towards a theoretical understanding. In *Beyond Farmer First*, ed. I. Scoones & J. Thompson, pp. 16–32. London: International Institute for Environment and Development.

Scott, J.D. (1956). The study of primordial buds and the reaction of roots to defoliation as the basis of grassland management. In *Proceedings VII International Grassland Congress*, pp. 479–87.

Seligman, W.G. (1993). Modelling as a tool for grassland science progress. In *Proceedings XVII International Grassland Congress*, pp. 743–8.

Serrão, E.A., Uhl, C. & Nepstad, D.C. (1993). Deforestation for pasture in the humid tropics: is it economically and environmentally sound in the long term? In *Proceedings XVII International Grassland Congress*, pp. 2215–21.

Shaw, N.H. (1978). Superphosphate and stocking rate effects on a native pasture oversown with *Stylosanthes humilis* in central coastal Queensland. 2. Animal production. *Australian Journal of Experimental Agriculture and Animal Husbandry*, **18**, 800–7.

Sheard, R.W. (1970). Characterization of food reserves as a basis for timing nitrogen applications for timothy (*Phleum pratense* L.). In *Proceedings XI International Grassland Congress*, pp. 570–4.

Shelton, H.M., Piggin, C.M. & Brewbaker, J.L. (ed.) (1995). *Leucaena –
Opportunities and Limitations*. Canberra: ACIAR Proceedings 57.
Shepherd, J.A. & Coombs, R.F. (1979). The effect of four *Meloidogyne* species
(Nematoda: Meloidogynidae) on *Panicum maximum* cv. Umtali. *Rhodesian
Journal of Agricultural Research*, **17**, 155–6.
Sheriff, D.W. & Kaye, P.E. (1977). Responses of diffusive conductance to
humidity in a drought avoiding and a drought resistant (in terms of stomatal
response) legume. *Annals of Botany*, **41**, 653–5.
Sibbald, A.R., Maxwell, T.J., Morgan, T.E.H., Jones, J.R. & Rees, M.E. (1994).
The implications of controlled grazing sward height for the operation and
productivity of upland sheep systems in the UK. 2. Effects of two annual
stocking rates in combination with two levels of fertilizer nitrogen. *Grass and
Forage Science*, **49**, 89–95.
Sibma, L. & Alberda, T. (1980). The effect of cutting frequency and nitrogen
fertilizer rates on dry matter production, nitrogen uptake and herbage nitrate
content. *Netherlands Journal of Agricultural Science*, **28**, 243–51.
Simonsen, O. (1985). Herbage breeding in northern areas. In *Plant Production in
the North*, ed. A. Kaurin, O. Junttila & J. Nilsen, pp. 227–95. Tromso,
Norway: Norwegian University Press.
Skovlin, J. (1987). Southern Africa's experience with intensive short duration
grazing. *Rangelands*, **9**, 162–7.
Smith, C.J., Freney, J.R. & Mosier, A.R. (1993). Effect of acetylene provided by
wax-coated calcium carbide on transformations of urea nitrogen applied to
an irrigated wheat crop. *Biology and Fertility of Soils*, **16**(12), 86–92.
Smith, D. (1950). Seasonal fluctuations of root reserves in red clover, *Trifolium
pratense* L. *Plant Physiology*, **25**, 702–10.
Smith, R.R. & Quesenberry, K.H. (1993). Characterization of US collection of
clover and special-purpose legume germplasm. In *Proceedings XVII
International Grassland Congress*, pp. 203–4.
Smythe, D.S. & Checkland, P.B. (1976). Using a systems approach: the structure
of root definitions. *Journal of Applied Systems Analysis*, **5**, 75–83.
Sollenberger, L.E. & Jones, C.S. (1989). Beef production from nitrogen-fertilized
Mott dwarf elephantgrass and Pensacola bahiagrass pastures. *Tropical
Grasslands*, **23**, 129–34.
Spears, J.W. (1994). Minerals in forages. In *Forage Quality, Evaluation and
Utilization*, ed. G.C. Fahey Jr, M. Collins, D.R. Mertens & L.E. Moser, pp.
281–317. Madison, WI: American Society of Agronomy.
Spedding, C.R.W. (1970). The relative complexity of grassland systems. In
Proceedings XI International Grassland Congress, pp. A 126–31.
Sprague, V.G. (1952). Maintaining legumes in the sward. In *Proceedings VI
International Grassland Congress*, pp. 443–9.
Stafford Smith, M., Campbell, B., Steffen, W., Archer, S. & Ojima, D. (ed.)
(1995). *Global Change Impacts on Pastures and Rangelands, Implementation
Plan*. Canberra: GCTE Core Project Office.
Stafford Smith, D.M. & Foran, B.D. (1990). RANGEPACK: the philosophy
underlying the development of a microcomputer-based decision support
system for pastoral land management. *Journal of Biogeography*, **17**, 541–6.
Stafford Smith, M., Ojima, D. & Carter, J. (1996). Integrated approaches to
assessing sequestration opportunities for carbon in rangelands. In *Combating
Global Warming by Combating Land Degradation*, ed. V.R. Squires. Nairobi:
UNEP (in press).
Stakelum, G. & Dillon, P. (1989). The effect of herbage mass on the herbal intake

and grazing behaviour of dairy cows. In *Proceedings XVI International Grassland Congress*, pp. 1157–8.

Stapledon, R.G. (1933). *University College of Wales, Aberystwyth Welsh Plant Breeding Station. An Account of the Organization and Work of the Station from its Foundation in April, 1919 to July, 1933.* Aberystwyth, UK: Welsh Plant Breeding Station.

Stapledon, R.G. (1937). Presidential address. In *Report IV International Grassland Congress*, pp. 1–6.

Stapledon, R.G. (1944). *The Land, Now and Tomorrow,* 3rd edn. London: Faber & Faber.

Stapledon, R.G. & Davies, W. (1930). Experiments to test the yield and other properties of various species and strains of herbage plants under different methods of management. Aberystwyth, UK: Welsh Plant Breeding Station Bulletin Series H No. 10.

Stebbins, G.L. (1952). Species hybrids in grasses. In *Proceedings VI International Grassland Congress*, pp. 247–53.

Steele, K.W. & Vallis, I. (1988). The nitrogen cycle in pastures. In *Advances in Nitrogen Cycling in Agricultural Ecosystems*, ed. J.R. Wilson, pp. 274–91. Wallingford, UK: CAB International.

Steinke, T.D. & Booysen, P. de V. (1968). The regrowth and utilization of carbohydrate reserves of *Eragrostis curvula* after different frequencies of defoliation. In *Proceedings of Grassland Society of Southern Africa*, 3, 105–10.

Stern, W.R., Gladstones, J.S., Francis, C.M., Collins, W.J., Chatel, D.L., Nicholas, D.A., Gillespie, D.J., Wolfe, E.C., Southwood, O.R., Beale, P.E. & Curnow, B.C. (1983). Subterranean clover improvement: an Australian program. In *Proceedings XIV International Grassland Congress*, pp. 116–19.

Stobbs, T.H. (1973a). The effect of plant structure on the intake of tropical pastures. 1. Variation in the bite size of grazing cattle. *Australian Journal of Agricultural Research*, 24, 809–19.

Stobbs, T.H. (1973b). The effect of plant structure on the intake of tropical pastures. 2. Differences in sward structure, nutritive value, and bite size of animals grazing *Setaria anceps* and *Chloris gayana* at various stages of growth. *Australian Journal of Agricultural Research*, 24, 821–9.

Stobbs, T.H. (1975). A comparison of Zulu sorghum, bulrush millet and white panicum in terms of yield, forage quality and milk production. *Australian Journal of Experimental Agriculture and Animal Husbandry*, 15, 211–18.

Stobbs, T.H. (1977). Seasonal changes in the preference by cattle for *Macroptilium atropurpureum* cv. Siratro. *Tropical Grasslands*, 11, 87–92.

Stobbs, T.H., Minson, D.J. & McLeod, M.N. (1977). The response of dairy cows grazing a nitrogen fertilized grass pasture to a supplement of protected protein. *Journal of Agricultural Science,* 89, 137–41.

Stoddart, L.A. & Smith, A.D. (1943). *Range Management.* New York: McGraw Hill.

Strange, R. (1952). Observations on the use of tropical legumes in tropical mixed farming. In *Proceedings VI International Grassland Congress*, pp. 1420–5.

Stuth, J.W., Hamilton, W.T., Conner, J.C. & Sheehy, D.P. (1993). Decision support systems in the transfer of grassland technology. In *Proceedings XVII International Grassland Congress*, pp. 749–56.

Stuth, J.W., Kirby, D.R. & Chmielewski, R.E. (1981). Effect of herbage allowance on the efficiency of defoliation by the grazing animal. *Grass and Forage Science*, 36, 9–15.

Suginobu, K., Takamizo, T., Komatsu, T., Akiyama, F. & Tominaga, Y. (1993). Intergeneric hybrids of male sterile Italian ryegrass crossed with tall fescue. In *Proceedings XVII International Grassland Congress*, pp. 436–7.

Sullivan, J.T. & Sprague, V.G. (1943). Composition of the roots and stubble of perennial ryegrass following partial defoliation. *Plant Physiology*, **18**, 656–70.

Suzuki, M. (1993). Fructans in crop production and preservation. In *Science and Technology of Fructans*, ed. M. Suzuki & N.J. Chatterton, pp. 227–55. Florida: CRC Press.

Tabe, L.M., Higgins, C.M., McNabb, W.C. & Higgins, T.J.V. (1993). Genetic engineering of grain and pasture legumes for improved nutritive value. *Genetica*, **90**, 181–200.

Tainton, N.M. (1981). *Veld and Pasture Management in South Africa*. Pietermaritzburg: Shuter & Shooter.

Tainton, N.M. (1985). Recent trends in grazing management philosophy in South Africa. *Journal of the Grassland Society of Southern Africa*, **2**, 4–6.

Takahashi, M. (1952). Tropical forage legumes and browse plants research in Hawaii. In *Proceedings VI International Grassland Congress*, pp. 1411–17.

Tans, P.P., Fung, I.Y. & Takahashi, T. (1990). Observational constraints on the global atmospheric CO_2 budget. *Science*, **247**, 1431–8.

Taylor, C.A., Garza, N.E. & Brooks, T.D. (1993). Grazing systems on the Edwards Plateau of Texas: are they worth the trouble? I. Soil and vegetation response. *Rangelands*, **15**, 53–60.

Taylorson, R.B. & Hendricks, S.B. (1977). Dormancy in seeds. *Annual Review of Plant Physiology*, **28**, 331–354.

Teitzel, J.K. (1992). Sustainable pasture systems in the humid tropics of Queensland. *Tropical Grasslands*, **26**, 196–205.

Teitzel, J.K., Gilbert, M.A. & Cowan, R.T. (1991). Sustaining productive pastures in the tropics. 6. Nitrogen fertilized grass pastures. *Tropical Grasslands*, **25**, 111–18.

Thomas, C. & Young, J.W.O. (ed.) (1982). *Milk from Grass*. Cleveland: ICI. Hurley, UK: Grassland Research Institute.

Thomas, R.J. (1992). The role of the legume in the nitrogen cycle of productive and sustainable pastures. *Grass and Forage Science*, **47**, 133–42.

Thomas, W.J. (1960). Some economic aspects of grassland production and utilization. In *Proceedings VIII International Grassland Congress*, pp. 27–34.

Thornley, J.H.M., Bergelson, J. & Parsons, A.J. (1995). Complex dynamics in a carbon–nitrogen model of a grass–legume pasture. *Annals of Botany*, **75**, 79–94.

Tilley, J.M.A. & Terry, R.A. (1963). A two-stage technique for *in vitro* digestion of forage crops. *Journal of British Grassland Society*, **18**, 104–11.

Tilman, D. (1988). *Plant Strategies and the Dynamics and Structure of Plant Communities*. Princeton: Princeton Monographs.

Timmerman, G.M. & McCallum, J.A. (1993). Applications of random amplified polymorphic DNA (RAPD) markers in plant biology. In *Proceedings XVII International Grassland Congress*, pp. 1025–31.

Titchen, N.M. & Scholefield, D. (1992). The potential of a rapid test for soil mineral nitrogen to determine tactical applications of fertilizer nitrogen to grassland. *Aspects of Applied Biology*, **30**, 223–9.

Togamura, Y., Ochiai, K. & Shioya, S. (1993). Quality of pasture managed to different leaf lengths. In *Proceedings XVII International Grassland Congress*, pp. 900–1.

Toledo, J.M. (1985). Pasture development for cattle production in the major ecosystems of the tropical American lowlands. In *Proceedings XV International Grassland Congress*, pp. 74–81.

Toledo, J.M. & Formoso, D. (1993). Sustainability of sown pastures in the tropics and subtropics. In *Proceedings XVII International Grassland Congress*, pp. 1891–6.

Toledo, J.M., Giraldo, H. & Spain, J.M. (1987). Effecto del pastoreo continuo y el métoda de siembra en la persistencia de la associación *Andropogon gayanus/Stylosanthes capitata*. *Pasturas Tropicales*, **9**(3), 18–24.

Trongkongsin, K. & Humphreys, L.R. (1988). The long–short day requirement for flowering in *Stylosanthes guianensis*. *Australian Journal of Agricultural Research*, **39**, 199–207.

Tsuiki, M., Takahashi, S. & Oku, T. (1993). Application of rural networks to the extraction of various types of grasslands in Japan using Landsat thematic mapper data. In *Proceedings XVII International Grassland Congress*, pp. 786–7.

Tutin, T.G. (1958). Classification of the legumes. In *Nutrition of the Legumes*, ed. E.G. Hallsworth, pp. 3–14. London: Butterworths Scientific Publications.

Ulyatt, M.J., Macrae, J.C., Clarke, R.T.J. & Pearce, P.D. (1975). Quantitative digestion of fresh herbage of sheep. IV. Protein synthesis in the stomach. *Journal of Agricultural Science*, **84**, 453–8.

UNEP (1991). *Status of Desertification and Implementation of the United Nations Plan of Action to Combat Desertification*. Nairobi: UNEP.

Vallis, I. & Gardener, C.J. (1984). Short-term nitrogen balance in urine-treated areas of pasture on a yellow earth in the subhumid tropics of Queensland. *Australian Journal of Experimental Agriculture and Animal Husbandry*, **24**, 522–8.

Vallis, I., Haydock, K.P., Ross, P.J. & Henzell, E.F. (1967). Isotopic studies on the uptake of nitrogen by pasture plants. III. The uptake of small additions of ^{15}N-labelled fertilizer by rhodes grass and townsville lucerne. *Australian Journal of Agricultural Research*, **18**, 865–77.

Vallis, I., Henzell, E.G. & Evans, T.R. (1977). Uptake of soil nitrogen by legumes in mixed swards. *Australian Journal of Agricultural Research*, **28**, 413–25.

Vallis, I. & Jones, R.J. (1973). Net mineralization of nitrogen in leaves and leaf litter of *Desmodium intortum* and *Phaseolus atropurpureus* mixed with soil. *Soil Biology and Biochemistry*, **5**, 391–8.

Vallis, I., Peake, (the late) D.C.I., Jones, R.K. & McCown, R.L. (1985). Fate of urea-nitrogen from cattle urine in a pasture–crop sequence in a seasonally dry tropical environment. *Australian Journal of Agricultural Research*, **36**, 809–17.

van der Kley, F.K. (1956). A simple method for the accurate estimation of daily variations in the quality and quantity of herbage consumed by rotationally grazed cattle and sheep. *Netherlands Journal of Agricultural Science*, **4**, 197–204.

Van der Molen, H. & t'Hart, M.L. (1966). Possibilities of increasing grassland production with nitrogen fertilizers in European agriculture. In *Proceedings X International Grassland Congress*, pp. 36–44.

van Dijk, G. & Hoogervorst, N. (1982). The demand for grassland in Europe towards 2000. Some implications of a possible scenario. In *Efficient Grassland Farming*, ed. A.J. Corrall, pp. 21–31. Hurley, UK: British Grassland Society Occasional Symposium 14.

Van Dyne, G.M. (1970). A systems approach to grasslands. In *Proceedings XI International Grassland Congress*, pp. A131–43.

Van Heerden, J.M. & Tainton, W.M. (1989). Development of a general relationship between stocking rate and animal production. In *Proceedings XVI International Grassland Congress*, pp. 1103–4.

Van Soest, P.J. (1967). Development of a comprehensive system of feed analysis and its application to forages. *Journal of Animal Science* 26, 119–28.

Van Wijk, A.J.P., Boonman, J.G. & Rumball, W. (1993). Achievements and perspectives in the breeding of forage grasses and legumes. In *Proceedings XVII International Grassland Congress*, pp. 379–83.

Vasil, I.K. (1985). Biotechnology in the improvement of forage crops. In *Proceedings XV International Grassland Congress*, pp. 45–8.

Vaughan, D. & Ord, B.G. (1985). Soil organic matter – a perspective of its nature, extraction, turnover and role in soil fertility. In *Soil Organic Matter and Biological Activity*, ed. D. Vaughan & R.E. Malcolm, pp. 1–35. Dordrecht: Martinus Nijhoff/Dr W. Junk Developments in Plant and Soil Sciences 16.

Vélez, C.A. & Escobar, L.G. (1970). Effects of seasonal N application to Paragrass on beef production. *Acta Agronomia*, 20, 65–90.

Vezzani, V. & Carbone, E. (1937). The influence of mountain pasturing on the development of young cattle. In *Report Fourth International Grassland Congress*, pp. 21–5. Aberystwyth, UK.

Vicente-Chandler, J., Silva, S. & Figarella, J. (1959). The effect of nitrogen fertilization and frequency of cutting on the yield and composition of three tropical grasses. *Agronomy Journal*, 51, 202–6.

Vieira, J.M. (1985). The effect of type of pasture and stocking rate on pasture characteristics, animal behaviour and production. PhD Thesis, University of Queensland.

Vincent, J.M. (1942). Serological studies of the root-nodule bacteria. II. Strains of *Rhizobium trifolii*. *Proceedings of Linnaean Society of New South Wales*, 67, 82–6.

Vincent, J.M. (1956). Strains of rhizobia in relation to clover establishment. In *Proceedings VII International Grassland Congress*, pp. 179–89.

Virtanen, A.I. (1937). Associated growth of legumes and non-legumes. In *Report IV International Grassland Congress*, pp. 78–89.

Voisin, A. (1959). *Grass Productivity*. London: Crosby Lockwood.

Volenec, J.J. (1986). Nonstructural carbohydrates in stem base components of tall fescue during regrowth. *Crop Science*, 26, 122–7.

Volesky, J.D., Lewis, J.K. & Butterfield, C.H. (1990). High-performance short-duration and repeated-seasonal grazing systems: effect on diets and performance of calves and lambs. *Journal of Range Management*, 43, 310–15.

Volio, G.C.A. (1952). Problems in development of a grassland program in the American tropics. In *Proceedings VI International Grassland Congress*, pp. 141–8.

Wade, M.H., Peyroud, J.L., Lemaire, G. & Cameron, E.A. (1989). The dynamics of daily area and depth of grazing and herbage intake of cows in a five day paddock system. In *Proceedings XVI International Grassland Congress*, pp. 1111–12.

Walker, B. (1980). Effects of stocking rate on perennial tropical legume grass pastures. PhD Thesis, University of Queensland.

Walker, B., Hodge, P.B. & O'Rourke, P.K. (1987). Effects of stocking rate and grass species on pasture and cattle productivity of sown pastures on a fertile brigalow soil in central Queensland. *Tropical Grasslands*, 21, 14–23.

Walker, B. & Weston, E.J. (1990). Pasture development in Queensland – a success story. *Tropical Grasslands*, 24, 257–68.

Walker, B.H. (1988). Autecology, synecology, climate and livestock as agents of rangeland dynamics. *Australian Rangelands Journal*, **10**, 69–75.

Walker, B.H. (1993). Stability in rangelands: ecology and economics. In *Proceedings XVII International Grassland Congress*, pp. 1885–90.

Walker, B.H., Ludwig, D., Holling, C.S. & Peterman, R.M. (1981). Stability of semi-arid savanna grazing systems. *Journal of Ecology*, **69**, 473–98.

Walker, T.W. (1956). Nitrogen and herbage production. In *Proceedings VII International Grassland Congress*, pp. 157–67.

Walker, T.W., Orchiston, H.D. & Adams, A.F.R. (1954). The nitrogen economy of grass legume associations. *Journal of the British Grassland Society*, **9**, 249–74.

Warmke, H.E. (1952). Apomixis in grasses, with special reference to *Panicum maximum*. In *Proceedings VI International Grassland Congress*, pp. 209–15.

Warren, S.D., Thurow, T.L., Blackburn, W.H. & Garza, W.E. (1986). The influence of livestock trampling under intensive rotation grazing on soil hydrologic characteristics. *Journal of Range Management*, **39**, 491–5.

Warren-Wilson, J. (1961). Influence of spatial arrangement of foliage area on light interception and pasture growth. In *Proceedings VIII International Grassland Congress*, pp. 275–9.

Waters-Bayer, A. & Waters-Bayer, W. (1994). *Planning with Pastoralists: PRA and More*. Göttingen: Deutsche Gesellschaft für Technische Zusammenarbeit.

Watkin, B.R. & Clements, R.J. (1978). The effects of grazing animals on pastures. In *Plant Relations in Pastures*, ed. J.R. Wilson, pp. 273–89. Melbourne: CSIRO.

Watson, D.J. (1947). Comparative physiological studies on the growth of field crops. 1. Variation in net assimilation rate and leaf area between species and varieties, and within and between years. *Annals of Botany*, **11**, 41–

Watson, D.J. (1952). The physiological basis of variation in yield. *Advances in Agronomy*, **4**, 101–45.

Watson, D.J. (1968). A prospect of crop physiology. *Annals of Applied Biology*, **62**, 1–9.

Watson, J. & Graves, J.D. (1993). Effect of elevated carbon dioxide on the performance of nine coexisting grassland species. In *Proceedings XVII International Grassland Congress*, pp. 1145–7.

Watson, S.E. & Whiteman, P.C. (1981). Grazing studies on the Guadalcanal Plains, Solomon Islands. 2. Effects of pasture mixtures and stocking rate on animal production and pasture components. *Journal of Agricultural Science*, **97**, 353–64.

Weaver, J.E. & Clements, F.E. (1938). *Plant Ecology*, 2nd edn. New York: McGraw-Hill.

Weinmann, H. (1948). Effects of grazing intensity and fertiliser treatment on Transvaal highveld. *Empire Journal of Experimental Agriculture*, **16**, 111–18.

Weinmann, H. (1949). Productivity of Marandellas sandveld pasture in relation to frequency of cutting. *Rhodesia Agricultural Journal*, **46**, 175–89.

Weinmann, H. (1952). Carbohydrate reserves in grasses. In *Proceedings VI International Grassland Congress*, pp. 655–60.

Weinmann, H. (1961). Total available carbohydrates in grasses and legumes. *Herbage Abstracts*, **31**, 255–61.

Welker, J.M., Rykiel, E.J. Jr, Briske, D.D. & Goeschl, J.D. (1985). Carbon import among vegetative tillers within two bunch grasses: assessment with carbon-11 labelling. *Oecologia*, **67**, 209–12.

Welz, M., Wood, M.K. & Parker, E.E. (1989). Flash grazing and trampling: effects on infiltration rates and sediment yield on a selected New Mexico range site. *Journal of Arid Environments*, **16**, 95–100.

West, S.H. (1970). Biochemical mechanism of photosynthesis and growth depression in *Digitaria decumbens* when exposed to low temperatures. In *Proceedings XI International Grassland Congress*, pp. 514–7.

Westoby, M., Walker, B.H. & Noy-Meir, I. (1989). Opportunistic management for rangelands not at equilibrium. *Journal of Range Management*, **42**, 266–74.

White, D.W.R., Biggs, D.R., McManus, M.T., Voisey, C.R., Christeller, J.T., Broadwell, A.,H., Burgess, E.J.P., Chilcott, C.N., Wigley, P.J. & McGregor, P.G. (1993). Development of plants resistant to insect pests using gene manipulation. In *Proceedings XVII International Grassland Congress*, pp. 1159–61.

Whitehead, D. (1995). *Grassland Nitrogen*. Wallingford, UK: CAB International.

Whiteman, P.C. (1970). Seasonal changes in growth and nodulation of perennial tropical pasture legumes in the field. II. Effects of controlled defoliation levels on the nodulation of *Desmodium intortum* and *Phaseolus atropurpureus*. *Australian Journal of Agricultural Research*, **21**, 207–14.

Whitney, A.S. & Green, R.E. (1969). Legume contributions to yields and compositions of *Desmodium* spp. – pangolagrass mixtures. *Agronomy Journal*, **61**, 741–6.

Whittaker, A.D. (1993). Decision support systems and expert systems for range science. In *Emerging Issues for Decision Support Systems for Resource Management*, ed. J.W. Stuth & B.G. Lyons, pp. 69–81. Carnforth, UK: UNESCO Man and the Biosphere Book 11, Parthenon Publishing.

Whittaker, R.H. (1953). A consideration of climax theory. The climax as a population and pattern. *Ecological Monographs*, **23**, 41–78.

Whyte, R.O. (1960). Grassland in a developing world. In *Proceedings VIII International Grassland Congress*, pp. 11–15.

Wilkins, R.J. (1993). Environmental constraints to production systems. In *The Place for Grass in Land Use Systems*, British Grassland Society Winter Meeting 1993, pp. 19–30. Reading, UK: British Grassland Society, University of Reading.

Wilkins, R.J., Garwood, E.A., Hopkins, A. & Tallowin, J.R.B. (1987). Beef production from permanent grassland in Britain. *Irish Grassland and Animal Production Association Journal*, **22**, 71–7.

Wilkins, R.J. & Harvey, H.J. (1994). Management options to achieve agricultural and nature conservation objectives. In *Grassland Management and Nature Conservation*, ed. R.J. Haggar & S. Peel, pp. 86–94. Reading, UK: British Grassland Society Occasional Symposium 28.

Wilkinson, P.R. (1964). Pasture spelling as a control measure for cattle ticks in southern Queensland. *Australian Journal of Agricultural Research*, **15**, 822–40.

Wilkinson, S.R. & Frere, M.H. (1993). Use of forage nitrate–nitrogen to improve nitrogen-use efficiency in coastal bermudagrass. In *Proceedings XVII International Grassland Congress*, pp. 1450–1.

Williams, C.H. (1980). Soil acidification under clover pasture. *Australian Journal of Experimental Agriculture and Animal Husbandry*, **20**, 561–7.

Williams, R.D. (1937). Genetics of red clover and its bearing on practical breeding. In *Report IV International Grassland Congress*, pp. 238–51.

Williams, R.J. (1983). Tropical legumes. In *Genetic Resources of Forage Plants*, ed. J.G. McIvor & R.A. Bray, pp. 17–37. Melbourne: CSIRO.

Williams, T.E. & Strange, L.R. (1952). Resumé of contributions in grassland production. In *Proceedings VI International Grassland Congress*, pp. 195–9.

Williams, W.A. (1967). The role of the Leguminosae in pasture and soil improvement in the neotropics. *Tropical Agriculture*, **44**, 103–15.

Williams, W.M. & Williams, E.G. (1983). Use of embryo culture with nurse endosperm for interspecific hybridization in pasture legumes. In *Proceedings XIV International Grassland Congress*, pp. 163–5.

Williams, W.T. & Burt, R.L. (1982). A re-appraisal of Hartley's agrostological index. *Journal of Applied Ecology*, **19**, 159–66.

Willoughby, W.M. (1959). Limitations to animal production imposed by seasonal fluctuations in pasture and by management procedures. *Australian Journal of Agricultural Research*, **10**, 248–68.

Wilm, H.G. (1952). The relation of different kinds of plant cover to water yields in semiarid areas. In *Proceedings VI International Grassland Congress*, pp. 1046–55.

Wilson, A.D. & Leigh, J.H. (1967). Comparison of the productivity of sheep grazing natural pastures of the Riverine Plain. *Australian Journal of Experimental Agriculture and Animal Husbandry*, **10**, 549–54.

Wilson, A.D. & Tupper, G.J. (1982). Concepts and factors applicable to the measurement of range condition. *Journal of Range Management*, **35**, 684–9.

Wilson, D. (1975). Variation in leaf respiration in relation to growth and photosynthesis of *Lolium*. *Annals of Applied Biology*, **80**, 323–38.

Wilson, J. (1988). *Changing Agriculture. An Introduction to Systems Thinking*. Kenthurst, Australia: Kangaroo Press.

Wilson, J.R., Akin, D.E., McLeod, M.N. & Minson, D.J. (1989). Particle size reduction of the leaves of a tropical and temperate grass by cattle. II. Relation of anatomical structure to the process of leaf breakdown through chewing and digestion. *Grass and Forage Science*, **44**, 65–75.

Wilson, J.R., Brown, R.H. & Windham, W.R. (1983). Influence of leaf anatomy on the dry matter digestibility of C_3, C_4 and C_3/C_4 intermediate types of *Panicum* species. *Crop Science*, **23**, 141–6.

Wilson, J.R. & Ford, C.W. (1971). Temperature influences on the growth, digestibility, and carbohydrate composition of two tropical grasses, *Panicum maximum* var. *trichoglume* and *Setaria sphacelata*, and two cultivars of the temperate grass *Lolium perenne*. *Australian Journal of Agricultural Research*, **22**, 563–71.

Wilson, J.R. & Hattersley, P.W. (1983). *In vitro* digestion of bundle sheath cells in rumen fluid and its relation to the suberized lamella and C_4 photosynthetic type in *Panicum* species. *Grass and Forage Science*, **38**, 219–23.

Wilson, J.R. & Haydock, K.P. (1971). The comparative response of tropical and temperate grasses to varying levels of nitrogen and phosphorus nutrition. *Australian Journal of Agricultural Research*, **22**, 573–87.

Wilson, J.R. & Ludlow, M.M. (1991). The environment and potential growth of herbage under plantations. In *Forages for Plantation Crops*, ed. H.M. Shelton & W.W. Stür, pp. 10–24. Canberra: ACIAR Proceedings 32.

Wilson, J.R. & Mannetje, L.'t. (1978). Senescence, digestibility and carbohydrate content of buffel grass and green panic leaves in swards. *Australian Journal of Agricultural Research*, **29**, 503–16.

Wilson, J.R. & Minson, D.J. (1980). Prospects for improving the digestibility and intake of tropical grasses. *Tropical Grasslands*, **14**, 253–9.

Wilson, K. & Morren, Jr, G.E.B. (1990). *Systems Approaches for Improvement in Agriculture and Resource Management*. New York: Macmillan.

Witty, J.F. & Minchin, F.R. (1988). Measurement of nitrogen fixation by the acetylene reduction assay; myths and mysteries. In *Nitrogen Fixation by Legumes in Mediterranean Agriculture*, ed. D.P. Beck & L.A. Materon, pp. 331–44. Dordrecht: Martinus Nijhoff.

Woledge, J. & Leafe, E.L. (1976). Single leaf and canopy photosynthesis in a ryegrass sward. *Annals of Botany*, **40**, 773–83.

Wolf, J. & Janssen, L.H.J.M. (1991). Effects of changing land use in the Netherlands on net carbon fixation. *Netherlands Journal of Agricultural Science*, **39**, 237–46.

Wong, C.C. (1991). Shade tolerance of tropical forages: a review. In *Forages for Plantation Crops*, ed. H.M. Shelton & W.W. Stür, pp. 64–69. Canberra: ACIAR Proceedings 32.

Wong, C.C. (1993). Growth and persistence of two *Paspalum* species to defoliation in shade. PhD thesis, University of Queensland.

Wong, C.C., Rahim, H. & Sharudin, M.A.Mohd. (1985). Shade tolerance potential of some tropical forages for integration with plantations. 1. Grasses. *MARDI Research Bulletin*, **13**, 225–47.

Wong, C.C., Sharudin, M.A.Mohd. & Rahim, H. (1985). Shade tolerance potential of some tropical forages for integration with plantations. 2. Legumes. *MARDI Research Bulletin*, **13**, 249–69.

Woodmansee, R.G. & Riebsame, W.E. (1993). Evaluating the effects of climate change on grasslands. In *Proceedings XVII International Grassland Congress*, pp. 1191–6.

Woodward, F.I., Thompson, G.B. & McKee, I.F. (1991). The effects of elevated concentrations of carbon dioxide on individual plants, populations, communities and ecosystems. *Annals of Botany*, **67**, 23–38.

World Commission on Environment and Development (1987). *Our Common Future*. Oxford: Oxford University Press.

Wright, I.A. & Whyte, T.K. (1989). Effects of sward surface height on the performance of continuously stocked spring-calving beef cows and their calves. *Grass and Forage Science*, **44**, 259–66.

Index

abiotic factors, 136, 142, 155
acceptability, 62, 127, 140
acid rain, 8
adaptation, 42, 129
adoption, 73, 181, 188
affinities, rhizobial, 87
age structure, 147
agricultural potential, 84
agrostological index, 58
aluminium, 83, 91
animal
    as banking system, 3
    needs, 176, 179
    output, 74, 156
    purchase and sale, 163
anthracnose, 46
apomixis, 30, 63, 65
atmospheric pollution, 8
autecology, 154
availability of pasture, 153, 164
avoidance, 44, 143
Azospirillum, 94

bacteria, free-living, 94
beef, 167
biodiversity, 5, 36, 50, 58, 137
biological control, 48
biosphere, 14, 26
biotic stress, 45
boron, 83, 91
botanical composition, 2, 129, 136, 154,
    156, 161, 177
brown-rib mutants, 33
buds, 113
buffering, 64

calcium, 78, 83, 86
callus initiation, 70
carbohydrate utilization, 115
carbon

dioxide, 9, 61
    enrichment, 17
    flux, 11, 13, 120
    nitrogen ratio, 21, 111
centres of origin, 58
change, 2, 142, 154, 200
chemicals
    deactivation, 4, 45
    hard, 72
chlorofluorocarbons, 9, 21
chlorophyll, 40
classification, 56
climate, 9, 43, 57, 58, 130, 162
climax, 130, 136
cobalt, 35
collections, 52
community
    acceptance, 72
    attitudes, 194
    viability, 201
compatibility, 36
competition, 23, 42, 80, 116, 143
conceptualization, 196, 201
conservation
    forage, 2, 79, 170
    germplasm, 50
    nature, 5, 50, 61
    resources, 1, 4, 129, 148, 154
constructivist communication, 198
continuity of forage supply, 2
contour hedges, 4
control functions, 185
copper, 83, 91
criteria of merit, 30, 49
crop protection, 4
    systems, 3, 4, 15, 26, 201
cutting, 108, 115, 175
cysteine, 68
cytogenetics, 56, 65
cytokinins, 123